TOPOLOGY

Yong-seung CHO

KYOWOO

Yong-seung CHO

He earned his Ph.D. at the University of Chicago in 1987. He joined Brandeis University in the U.S., Kyungpook National University in Daegu, and Ewha-Womans University in Seoul as a professor of mathematics. He served as President of the Korean Mathematical Society (2003-2004), and played a fundamental role in establishing the National Institute for Mathematical Sciences (NIMS) and served as its first President in 2005-2008. His research specialties include Yang-Mills Theory, Seiberg-Witten invariants in 4-manifolds, and Gromov-Witten invariants and quantum cohomology as well as cosymplectic manifolds. In these fields, he has published more than 130 research papers and 30 books. He is currently Professor Emeritus at EwhaWomans University, and Invited Professor at Sungkyunkwan University.

Kyowoo Co., Ltd.
8, Yangnyeongsi-ro, Dongdaemun-gu,
Seoul, Korea 130-861,

Phone: 82-2-925-2861
F a x: 82-2-925-2860
e-mail: kyowoo@kyowoo.co.kr
http://www.kyowoo.co.kr

Printed in the Republic of Korea.
ISBN 979-11-251-0280-9 (93410)

₩ 22,000

Preface

A topological space is a set with a topological structure. The topology is of interest in its own right, that is, point set topology, algebraic topology, analytic topology and geometric topology. Also it is applied to natural sciences, especially to engineering mathematics, biomathematics, and physics. This book is an extended version of my expository lecture notes given at Sungkyunkwan University. We write this book for international students who need a book written by English. For the purpose we use our book Topology written in Korean and add simplicial homology, cohomology and de Rham isomorphism theorem.

At first, we introduce sets and functions, topological spaces, continuous maps, product spaces and quotient spaces, connectedness and compactness, classification of topological spaces, compactifications, paracompactness and so on.

Secondly, we deal with fundamental groups and covering spaces, differentiable manifolds and their properties, de Rham cohomology, simplicial homology and cohomology, and de Rham theorem : de Rham cohomologies are isomorphic to simplicial cohomologies with real coefficients for compact differentiable manifolds.

I would like to express my sincere thanks to CEO of Kyowoo Publishing

Co. Mr. Oh, Pan-Geoun and his amazing team for publishing this book and the greatest cover design I could ever imagine. This book was written at Ewha Womans University library. I would like to give special thanks to Sungkyunkwan University and Ewha Womans University for providing me special opportunities of lectures and spaces. My thanks also goes to Dr. Hyunjin Lee for her generous help with careful typing of the manuscript. Finally I want to thank my wife, Dr. Su Nam Kim for her inspiration and ever-patient proofreading. She was as important to this book getting done as I was. Thank you so much everyone!

Seoul, Republic of Korea

January 2019

Yong Seung Cho

Contents

Chapter 1

Sets and Maps

1.1 Sets

Recall a series of definitions on sets.

Definition 1.1.1. (1) Intuitively, a **set** is a collection of elements.

(2) An element x belongs to a set X, $x \in X$.

(3) x does not belong to X, $x \notin X$.

(4) A is a **subset** of X if every element of A is also an element of X, $A \subset X$.

(5) $X = Y$ if $X \subset Y$ and $Y \subset X$.

(6) The **union** $X \cup Y := \{x \mid x \in X \text{ or } x \in Y\}$.

(7) The **intersection** $X \cap Y := \{x \mid x \in X \text{ and } x \in Y\}$.

(8) The **empty set**, $\emptyset$, is the set having no element.

(9) X and Y are **disjoint** if $X \cap Y = \emptyset$.

(10) The **difference** $X - Y = \{x \mid x \in X \text{ and } x \notin Y\}$.

(11) $X \times Y = \{(x, y) \mid x \in X \text{ and } y \in Y\}$ is called the **(cartesian) product** of X and Y.

Remark 1.1.2 (Russell's Paradox). Does $M = \{A \mid A \notin A\}$ is a set? No! Since $M \in M$ iff $M \notin M$.

Definition 1.1.3. (1) Let $\{X_\alpha \mid \alpha \in I\}$ be a family of sets. The **product set** of the family is the set

$$\begin{aligned}\prod_{\alpha \in I} X_\alpha &= \{(x_\alpha)_{\alpha \in I} \mid x_\alpha \in X_\alpha \text{ for each } \alpha \in I\} \\ &= \{x : I \longrightarrow \bigcup_{\alpha \in I} X_\alpha \mid x(\alpha) \in X_\alpha \text{ for each } \alpha \in I\}.\end{aligned}$$

(2) A pair $(X, \leq)$ is a **partially ordered set** if X is a set with an order $\leq$ satisfying for all x, y, $z \in X$,

(i) $x \leq x$,

(ii) if $x \leq y$ and $y \leq x$, then $x = y$,

(iii) if $x \leq y$ and $y \leq z$, then $x \leq z$.

(3) A partially ordered set $(X, \leq)$ is a **totally ordered set** if for any x, $y \in X$, either $x \leq y$ or $y \leq x$, or $x = y$.

(4) Let $(X, \leq)$ be a partially ordered set.

(a) $x_0 \in X$ is a **maximal element** if $x \in X$, $x_0 \leq x$, then $x = x_0$.

(b) $x_0 \in X$ is a **minimal element** if $x \in X$, $x \leq x_0$, then $x = x_0$.

(c) $A \subset X$ is **bounded above** if there is an element $x \in X$ such that $a \leq x$ for all $a \in A$.

Lemma 1.1.4 (Zorn's Lemma). *Let $(X, \leq)$ be a partially ordered set. If every totally ordered subset of X has an upper bound in X, then X has a maximal element.*

Definition 1.1.5. (1) A relation $\sim$ on a set X is an **equivalence relation** on X if the relation satisfies the followings:

(i) $x \sim x$ for all $x \in X$,

(ii) if $x \sim y$, x, $y \in X$, then $y \sim x$,

(iii) if $x \sim y$ and $y \sim z$, x, y, $z \in X$, then $x \sim z$.

(2) The set $[x] := \{y \in X \mid x \sim y\}$ is called the **equivalence class** of x.

(3) The set $X/\sim := \big\{[x] \mid x \in X\big\}$ is called a **quotient set**.

Theorem 1.1.6. *Let $\sim$ be an equivalence relation on X.*

(1) *For each x, $y \in X$, either $[x] = [y]$ or $[x] \cap [y] = \emptyset$.*

(2) *$X/\sim = \{[x] \mid x \in X\}$ and $X = \cup_{[x] \in X/\sim} [x]$ is a partition of X.*

Definition 1.1.7. A **partition** of a set X is a collection of disjoint nonempty subsets of X whose union is all of X.

1.1. Exercises

1. Let A, B, $C \subset X$. Prove:

(1) $(A \cup B)^c = A^c \cap B^c$

(2) $(A \cap B)^c = A^c \cup B^c$

(3) $A \cap (B \cup C) = (A \cap B) \cup (A \cap C)$

(4) $A \times (B \cap C) = (A \times B) \cap (A \times C)$

2. Let $\sim$ be the relation in $\mathbb{R}$ defined by $a \sim b$ if and only if $a - b \in \mathbb{Z}$.

(1) Prove that $\sim$ is an equivalence relation.

(2) Find the equivalence class $[0]$.

(3) Find the quotient set $\mathbb{R}/\sim$.

3. Show that

$$\bigcap_{n=1}^{\infty}\left(-\frac{1}{n}, 2+\frac{1}{n}\right) = \bigcap_{n=1}^{\infty}\left[-\frac{1}{n}, 2+\frac{1}{n}\right] = \left[0, 2\right].$$

4. Let $\{X_\alpha \mid \alpha \in I\}$ be a family of nonempty sets. For each $\alpha \in I$, let A_α, B_α be subsets of X_α. Prove

(1) $\prod_{\alpha \in I} A_\alpha \bigcap \prod_{\alpha \in I} B_\alpha = \prod_{\alpha \in I}(A_\alpha \cap B_\alpha)$

(2) $\prod_{\alpha \in I} A_\alpha \bigcup \prod_{\alpha \in I} B_\alpha = \prod_{\alpha \in I}(A_\alpha \cup B_\alpha)$

1.2 Maps

Definition 1.2.1. Let X and Y be sets.

(1) A **map** $f : X \longrightarrow Y$ is a rule which assigns a unique element $f(x) \in Y$ for each element $x \in X$. Then

(i) X is called the domain of f,

(ii) Y is called the codomain of f,

(iii) $f(X) = \{f(x) \in Y \mid x \in X\}$ is called the image of f.

(2) Let $f, g : X \longrightarrow Y$ be two maps. Then they are the same :

$$f = g \quad \text{if} \quad f(x) = g(x) \ \text{ for all } x \in X.$$

Definition 1.2.2. Let $f : X \longrightarrow Y$ be a map.

(i) f is **injective** or **1-1** or **one-to-one**

$$\text{if } \ f(x_1) = f(x_2), \ \text{ then } \ x_1 = x_2.$$

(ii) f is **surjective** or **onto**

$$\text{if } \ f(X) = Y.$$

(iii) f is **bijective** if f is injective and surjective.

(iv) X and Y have the **same cardinality** if there is a bijective map $f : X \longrightarrow Y$.

(v) X is **countable** if either X is finite or there is a bijective map $f : \mathbb{N} \to X$.

(vi) X is **uncountable** if X is not countable.

(vii) Given maps $X \xrightarrow{f} Y$, $Y \xrightarrow{g} Z$ the **composite** of f and g is the map $g \circ f : X \longrightarrow Z$ defined by $(g \circ f)(x) := g(f(x))$ for all $x \in X$.

(viii) Let $f : X \longrightarrow Y$ be a map, $A \subset X$. Then the map $f|_A : A \longrightarrow Y$ given by $f|_A(a) = f(a)$, for all $a \in A$, is called the **restriction** of f.

(ix) If $f : X \to Y$ is a map and $B \subset Y$, then

$$f^{-1}(B) := \{x \in X \mid f(x) \in B\}$$

is called the **preimage** of B under f.

Example 1.2.3. (1) The finite set $\mathbb{Z}_n = \{0, 1, 2, \cdots, n-1\}$, and $\mathbb{Z}$, $\mathbb{N}$, $\mathbb{Q}$ are countable.

(2) $[0,1]$, $\mathbb{R}$, $\mathbb{C}$, and $D^2 = \{(x,y) \in \mathbb{R}^2 : x^2 + y^2 \leq 1\}$ are uncountable.

Axiom of Choice 1.2.4. *Let $\{X_\alpha \mid \alpha \in I\}$ be a family of nonempty sets X_α. Then there is a map $f : I \longrightarrow \cup_{\alpha\in I} X_\alpha$ such that $f(\alpha) \in X_\alpha$ for each $\alpha \in I$. That is, $\prod_{\alpha\in I} X_\alpha \neq \emptyset$.*

1.2. Exercises

1. Let $f : I \longrightarrow Y$ be a map.

(1) For subsets A and B of X,

(i) $f(A \cup B) = f(A) \cup f(B)$

(ii) $f(A \cap B) \subset f(A) \cap f(B)$

(iii) $f(A - B) \supset f(A) - f(B)$.

(2) For subsets C and D of Y,

(i) $f^{-1}(C \cup D) = f^{-1}(C) \cup f^{-1}(D)$

(ii) $f^{-1}(C \cap D) = f^{-1}(C) \cap f^{-1}(D)$

(iii) $f^{-1}(C - D) = f^{-1}(C) - f^{-1}(D)$.

(3) (i) $A \subset f^{-1}[f(A)]$ and equality holds if f is injective.

(ii) $C \supset f[f^{-1}(C)]$ and equality holds if f is surjective.

2. Let $f : (-1, 1) \longrightarrow \mathbb{R}$ be defined by $f(x) = \dfrac{x}{1 - |x|}$.

(1) Show that f is bijective.

(2) Find the inverse f^{-1} of f.

Chapter 2

Topological Spaces

2.1 Topology

Definition 2.1.1. (1) A **topology** on a set X is a collection $\mathcal{T}$ of subsets of X satisfying:

(i) $\emptyset \in \mathcal{T}$ and $X \in \mathcal{T}$,

(ii) The union of elements of any subcollection of $\mathcal{T}$ is in $\mathcal{T}$, i.e., for any subset $\mathcal{T}' \subset \mathcal{T}$, the union $\bigcup_{U \in \mathcal{T}'} U \in \mathcal{T}$,

(iii) The intersection of elements of any finite subcollection of $\mathcal{T}$ is in $\mathcal{T}$, i.e., for any finite subcollection $\mathcal{T}' \subset \mathcal{T}$, the intersection $\bigcap_{U \in \mathcal{T}'} U \in \mathcal{T}$.

(2) A set X with a topology $\mathcal{T}$, $(X, \mathcal{T})$, is called a **topological space**. An element $U \in \mathcal{T}$ is called an **open** set of X. The set $X - U$, $U \in \mathcal{T}$, is called a **closed** set of X.

Remark 2.1.2. For a topological space $(X, \mathcal{T})$

(1) $\emptyset$, X, arbitrary unions and finite intersections of open sets are open.

(2) X and $\emptyset$ are closed and open.

(3) For simplicity, we often say a space X instead of a topological space $(X, \mathcal{T})$.

Example 2.1.3. (1) For a set $X = \{a, b\}$, $\mathcal{P} = \{\emptyset, X, \{a\}, \{b\}\}$, $\mathcal{T}_1 = \{\emptyset, X\}$, $\mathcal{T}_2 = \{\emptyset, X, \{b\}\}$, $\mathcal{T}_3 = \{\emptyset, X, \{a\}\}$ are topologies of X.

(2) Let $X = \{a, b, c\}$. Then the collection of the subsets of X is

$$\mathcal{P} = \{\emptyset, X, \{a\}, \{b\}, \{c\}, \{a, b\}, \{a, c\}, \{b, c\}\}.$$

$\mathcal{T}_1 = \{\emptyset, X\}$, $\mathcal{T}_2 = \mathcal{P}$ are topologies of X. But $\mathfrak{F}_1 = \{\emptyset, X, \{a\}, \{b\}\}$, $\mathfrak{F}_2 = \{X, \emptyset, \{a, b\}, \{b, c\}\}$ are not topologies of X.

Definitions and Examples 2.1.4. For a nonempty set X,

(1) $\mathcal{T} = \{X, \emptyset\}$ is called the **indiscrete** topology of X, the $(X, \mathcal{T})$ is called a **indiscrete space**.

(2) $\mathfrak{D} = \{U \,|\, U \subset X\}$ is the set of all subsets of X is called the **discrete** topology, and $(X, \mathfrak{D})$ is called a **discrete space**.

(3) Let $d : X \times X \longrightarrow \mathbb{R}$ be a map such that

   (i) $d(x, y) \geq 0, \quad \forall x, \forall y \in X,$

   (ii) $d(x, y) = 0$ iff $x = y, \quad \forall x \in X,$

   (iii) $d(x, y) = d(y, x), \quad \forall x, \forall y \in X,$

   (iv) $d(x, z) \leq d(x, y) + d(y, z), \quad \forall x, \forall y, \forall z \in X.$

Then d is called a **metric** on X, and $d(x, y)$ is called the **distance** between x and y. For each point $x \in X$, $\forall \varepsilon > 0$,

$$B(x, \varepsilon) := \{y \in X \mid d(x, y) < \varepsilon\}$$

is called the **ε-neighborhood** of x. Moreover,

$$\mathcal{T}_d := \{U \subset X \mid \forall x \in U, \text{ there is } \varepsilon > 0 \text{ such that } B(x, \varepsilon) \subset U\}$$

is a topology for X (called a)**metric topology**).
The pair $(X, \mathcal{T}_d) := (X, d)$ is called a **metric space**.

(4) For $\mathbb{R}^n$, let $d(x, y) := |x - y|$. Then $(\mathbb{R}^n, d)$ is a metric space. $(\mathbb{R}^n, d)$ is called the **Euclidean space** of dimension n.

(5) Let $(X, \mathcal{T})$ be a topological space and $x \in X$. Then

$$\mathfrak{N}_x = \{U \in \mathcal{T} \mid x \in U\}$$

is called the **neighborhood system** of x.

Theorem 2.1.5. *For the neighborhood system $\mathfrak{N}_x$ of x in a topological space $(X, \mathcal{T})$,*

(1) *if $U \in \mathfrak{N}_x$, then $x \in U$.*

(2) *if $U \in \mathfrak{N}_x$, $U \subset V$, $V \in \mathcal{T}$, then $V \in \mathfrak{N}_x$.*

(3) *if U, $V \in \mathfrak{N}_x$, then $U \cap V \in \mathfrak{N}_x$.*

(4) *$U \in \mathcal{T}$ iff $\forall y \in U$, $U \in \mathfrak{N}_y$.*

Remark 2.1.6. (1) Let $(X, \mathcal{T}_1)$, $(X, \mathcal{T}_2)$ be two topological spaces.

$$\begin{aligned} \mathcal{T}_1 \subset \mathcal{T}_2 \text{ iff } & \mathcal{T}_1 \text{ is } \textit{weaker} \text{ (coarser) than } \mathcal{T}_2, \\ \text{iff } & \mathcal{T}_2 \text{ is } \textit{stronger} \text{ (finer) than } \mathcal{T}_1. \end{aligned}$$

Thus the indiscrete space $(X, \mathcal{T})$ is weaker than the discrete space $(X, \mathfrak{D})$.

(2) For the metric space $(\mathbb{R}^1, d)$,

$$\bigcap_{n=1}^{\infty} \Big(-2-\frac{1}{n},\, 2+\frac{1}{n}\Big) = [-2, 2],$$

$$\bigcup_{n=1}^{\infty} \Big[-2+\frac{1}{n},\, 2-\frac{1}{n}\Big] = (-2, 2).$$

In general, infinite intersection of open sets may not be open, infinite union of closed sets may not be closed.

(3) Let $(X, \mathcal{T})$ be a topological space, and $U_i \in \mathcal{T}$, $i = 1, 2, \cdots, n$, $U_\alpha \in \mathcal{T}$. Then

(i) $X - (\bigcup_\alpha U_\alpha) = \bigcap_\alpha (X - U_\alpha)$

: The intersection of any family of closed sets is closed.

(ii) $X - (\bigcap_{i=1}^{n} U_i) = \bigcup_{i=1}^{n} (X - U_i)$

: The union of any finite family of closed sets is closed.

2.1. Exercises

1. List all topologies on a set $X = \{x, y, z\}$ which consist of exactly four elements.

2. Let $f : X \longrightarrow Y$ be a map between sets X and Y.

(1) If $(Y, \mathcal{T}_Y)$ is a topological space, then show that

$$\mathcal{T}_X = \{f^{-1}(V) \mid V \in \mathcal{T}_Y\}$$

is a topology on X.

(2) If $(X, \mathcal{T}_X)$ is a topological space, then show that

$$\mathcal{T}_Y = \{V \subset Y \mid f^{-1}(V) \in \mathcal{T}_X\}$$

is a topology on Y.

3. Let $(\mathbb{R}, \mathcal{T}_d)$ be the usual real line, and $\mathcal{T} = \{\mathbb{R}, \emptyset\} \cup \{(a, \infty) \mid a \in \mathbb{R}\}$.

(1) Show that $\mathcal{T}$ is a topology on $\mathbb{R}$.

(2) Show that $\mathcal{T}_d \supsetneqq \mathcal{T}$.

(3) Let $\mathbb{Z}$ be the set of integers.

(i) Find the topology on $\mathbb{Z}$ as a subspace of $(\mathbb{R}, \mathcal{T}_d)$.

(ii) Find the topology on $\mathbb{Z}$ as a subspace of $(\mathbb{R}, \mathcal{T})$.

See (2.2.2) for the definition of subspace.

4. Let $(X, \mathcal{T}_d)$ be a metric space.
Define $d_1 : X \times X \to \mathbb{R}$ by $d_1(x, y) = \min\{d(x, y), 1\}$.

(1) Show that d_1 is a metric on X.

(2) Show that $\mathcal{T}_d = \mathcal{T}_{d_1}$.

2.2 Closure

Theorem 2.2.1. *Let $(X, \mathcal{T})$ be a topological space.*

(1) *X and $\emptyset$ are closed.*

(2) *Arbitrary intersections of closed sets are closed.*

(3) *Finite unions of closed sets are closed.*

Proof. Recall $F \subset X$ is closed iff $X - F$ is open, $U \subset X$ is open iff $X - U$ is closed.

(1) $X = X - \emptyset$, $X - X = \emptyset$.

(2) $X - (\bigcap_{\alpha} F_\alpha) = \bigcup_{\alpha}(X - F_\alpha)$.

(3) $X - (\bigcup_{i=1}^{n} F_i) = \bigcap_{i=1}^{n}(X - F_i)$.

□

Definition 2.2.2. Let $(X, \mathcal{T})$ be a topological space and $A \subset X$.

(1) Let $\mathcal{T}_A := \{A \cap U \,|\, U \in \mathcal{T}\}$. Then $\mathcal{T}_A$ is a topology of A. $(A, \mathcal{T}_A)$ is called a **subspace** of $(X, \mathcal{T})$.

(2) For $x \in X$, if any neighborhood U of x satisfies $A \cap (U - \{x\}) \neq \emptyset$, then x is called an **accumulation** (or a **limit point**) of A.
And the set $A' := \{x \in X \,|\, x \text{ is a limit point of } A\}$ is called the **derived set** of A.

Example 2.2.3. (1) If $A = [\,0, 1\,) \subset \mathbb{R}$, then $A' = [\,0, 1\,]$.

(2) If $B = [\,0, 1\,) \cup \{2\}$, then $B' = [\,0, 1\,]$.

(3) If $C = \{\frac{1}{n} \,|\, n \in \mathbb{N}\}$, then $C' = \{0\}$.

(4) If $D =$ 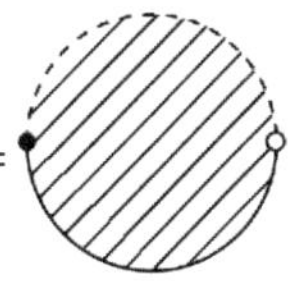 $\subset \mathbb{R}^2$, then $D' =$ 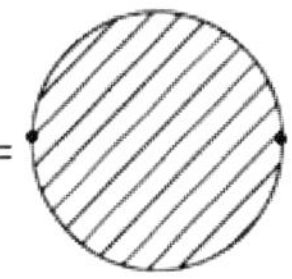

Definition 2.2.4. Let X be a topological space and $A \subset X$. Then the set $\bar{A} := A \cup A'$ is called the **closure** of A in X.

Theorem 2.2.5. *A is closed in X iff $A = \bar{A}$.*

Proof. A is closed iff $X - A$ is open.

(*) iff $A' \cap (X - A) = \emptyset$

iff $A' \subset A$

iff $\bar{A} = A \cup A' = A$.

Now prove $(*)$.

$(\Rightarrow)$ If $x \in X - A$, then since $X - A$ open,

$$\Big[(X - A) - \{x\}\Big] \cap A \subset (X - A) \cap A = \emptyset, \quad x \notin A'.$$

So, $A' \cap (X - A) = \emptyset$.

$(\Leftarrow)$ Let $A' \cap (X - A) = \emptyset$. If $x \in X - A$, then $x \notin A'$. There is a nbd $U(x)$ of x such that $A \cap \big[U(x) - \{x\}\big] = \emptyset$. Since $x \notin A$, $A \cap U(x) = \emptyset$ and $U(x) \subset X - A$. Thus

$$X - A = \bigcup_{x \in X - A} U(x)$$

is open. □

Corollary 2.2.6. *Let $(X, \mathcal{T})$ be a topological space, $A \subset X$, and $\mathfrak{F} = \{F \mid A \subset F \subset X,\ F$ is closed in $X\}$. Then*

$$\bar{A} = \bigcap_{F \in \mathfrak{F}} F.$$

Proof. "$\supset$" Since $A \subset \bar{A} = A \cup A'$, and by (2.2.1) $\bar{A}$ is closed. Since $\bar{A} \in \mathfrak{F}$, $\bar{A} \supset \bigcap_{F \in \mathfrak{F}} F$.

"$\subset$" If $F \subset \mathfrak{F}$, then $A \subset F$ and F closed. Show $A' \subset F$: then $\bar{A} \subset F$ and $\bar{A} \subset \bigcap_{F \in \mathfrak{F}} F$.

If $x \in A'$ and $x \notin F$, then $x \in X - F := U$ open, and $x \notin A$. So,

$$A \cap (U - \{x\}) \subset A \cap U \subset F \cap U = \emptyset,$$

and $x \notin A'$. Thus $\bar{A} = \bigcap_{F \in \mathfrak{F}} F$. □

Theorem 2.2.7. *Let X be a topological space, and A, $B \subset X$. Then*

(1) $\bar{\emptyset} = \emptyset$.

(2) $A \subset \bar{A}$.

(3) $\bar{\bar{A}} = \bar{A}$.

(4) $\overline{A \cup B} = \bar{A} \cup \bar{B}$.

(5) *If $\mathfrak{F} = \{A \subset X \mid A = \bar{A}\}$, then*

$$\mathcal{T} = \{U \subset X \mid U = X - F,\ F \in \mathfrak{F}\}$$

is the topology on X.

Proof. (5) (i) Since $\emptyset = \bar{\emptyset}$ and $X = \bar{X}$, $\emptyset$, $X \in \mathfrak{F}$. Thus

$$X = X - \emptyset \text{ and } \emptyset = X - X \in \mathcal{T}.$$

(ii) If $\{U_\alpha\}_{\alpha \in I} \subset \mathcal{T}$, then for each $\alpha \in I$, $X - U_\alpha := U_\alpha^c \in \mathfrak{F}$. By Theorem 2.2.1 (2),

$$\bigcap_{\alpha \in I} U_\alpha^c = \Big(\bigcup_{\alpha \in I} U_\alpha \Big)^c \in \mathfrak{F},$$

and thus $\bigcup_{\alpha \in I} U_\alpha \in \mathcal{T}$.

(iii) If $\{U_1, \cdots, U_n\} \subset \mathcal{T}$, then $\{U_1^c, \cdots, U_n^c\} \subset \mathfrak{F}$.
By Theorem 2.2.1 (3),

$$\bigcup_{i=1}^{n} U_i^c = \left(\bigcap_{i=1}^{n} U_i\right)^c \in \mathfrak{F}.$$

Thus $\bigcap_{i=1}^{n} U_i \in \mathcal{T}$.

□

Definition 2.2.8. Let X be a topological space, and $A \subset X$. A is said **dense** in X if $\bar{A} = X$.

Example 2.2.9. Let $(\mathbb{R}, d)$ be the real line, $\mathbb{Z} \subset \mathbb{Q} \subset \mathbb{R}$. Then

(1) $\mathbb{Z}' = \emptyset$, $\mathbb{Q}' = \mathbb{R}$, $\bar{\mathbb{Z}} = \mathbb{Z}$, $\bar{\mathbb{Q}} = \mathbb{R}$.

(2) $\bar{\mathbb{Q}} = \overline{\mathbb{R} - \mathbb{Q}} = \overline{\mathbb{R} - \mathbb{Z}} = \mathbb{R}$.

Thus $\mathbb{Q}$, $\mathbb{R} - \mathbb{Q}$, $\mathbb{R} - \mathbb{Z}$ are dense in $\mathbb{R}$, and $\mathbb{Z}$ is closed, but $\mathbb{Q}$ is not closed.

2.2. Exercises

1. Let $\mathbb{N}$ be the set of all natural numbers, and

$$\mathcal{T} = \{\emptyset\} \cup \{U_n \mid U_n = \{n, n+1, \cdots\}, n \in \mathbb{N}\}.$$

(1) Show that $\mathcal{T}$ is a topology on $\mathbb{N}$.

(2) List the open sets containing 4.

(3) Find the set of accumulation points of $\{2, 6, 10\}$.

(4) Find the subset A of $\mathbb{N}$ such that $\bar{A} = \mathbb{N}$.

2. Let A and B be subsets of a topological space $(X, \mathcal{T})$. Then show

(1) $(A \cup B)' = A' \cup B'$

(2) $\overline{(A \cup B)} = \bar{A} \cup \bar{B}$

(3) If $\{A_\alpha \mid \alpha \in I\}$ is a family of subsets in X, then show

(i) $\bigcup_{\alpha \in I} \overline{A_\alpha} \subset \overline{\bigcup_{\alpha \in I} A_\alpha}$

(ii) $\bigcap_{\alpha \in I} \overline{A_\alpha} = \overline{\bigcap_{\alpha \in I} A_\alpha}$

(iii) $\bigcup_{\alpha \in I} \overline{A_\alpha} = \overline{\bigcup_{\alpha \in I} A_\alpha}$ if $\bigcup_{\alpha \in I} \overline{A_\alpha}$ is closed.

3. Let $(\mathbb{R}, \mathcal{T}_d)$ be the real line. Let $A_n = (\frac{1}{n}, \infty)$, $n = 1, 2, \cdots$. Then

(1) find $\bigcup_{n=1}^{\infty} \overline{A_n}$,

(2) find $\overline{\bigcup_{n=1}^{\infty} A_n}$.

2.3 Interior, Exterior and Boundary

Definition 2.3.1. Let X be a topological space, and $A \subset X$.

(1) $i(A) := \bigcup_{\substack{U \subset A \\ U:\text{ open}}} U$ is called the **interior** of A, $x \in i(A)$ is called an **interior point** of A.

(2) $e(A) := i(X - A)$ is called the **exterior** of A, $x \in e(A)$ is called an **exterior point** of A.

(3) $b(A) := X - \big(i(A) \cup e(A)\big)$ is called the **boundary** of A. A point $x \in b(A)$ is called a **boundary point** of A.

Remark 2.3.2. $X = i(A) \cup e(A) \cup b(A)$ is a disjoint union.

Theorem 2.3.3. *Let X be a space, and $A \subset X$.*

(1) *A is open iff $A = i(A)$ iff $A \cap b(A) = \emptyset$.*

(2) *A is closed iff $A = \bar{A}$ iff $A \cup b(A) = A$ iff $b(A) \subset A$.*

Proof. Since $i(A) \subset A$, $e(A) \subset X - A$ and $b(A) = X - \big(i(A) \cup e(A)\big)$. Thus $X = i(A) \cup e(A) \cup b(A)$ is a disjoint union, and $i(A)$, $e(A)$ are open, $b(A)$ closed.

(1) Clear.

(2) A is closed iff $A = \bar{A}$

$$\begin{aligned} &\text{iff } X - A \text{ is open} \\ &\text{iff } X - A = i(X - A) = e(A). \end{aligned}$$

Then $A = X - (X - A) = X - e(A) = i(A) \cup b(A)$. Thus

$$b(A) \subset A \ \text{ iff } \ A = A \cup b(A).$$

Conversely, if $A = A \cup b(A)$, then $b(A) \subset A$. If $x \in X - A$, then $x \notin A$ and $x \notin b(A)$. The $x \in e(A) \subset X - A$. Thus $X - A$ is open and A closed.

□

Example 2.3.4. Consider the following metric spaces :

$$(\mathbb{R}, d_1) \hookrightarrow (\mathbb{R}^2, d_2) \hookrightarrow (\mathbb{R}^3, d_3)$$

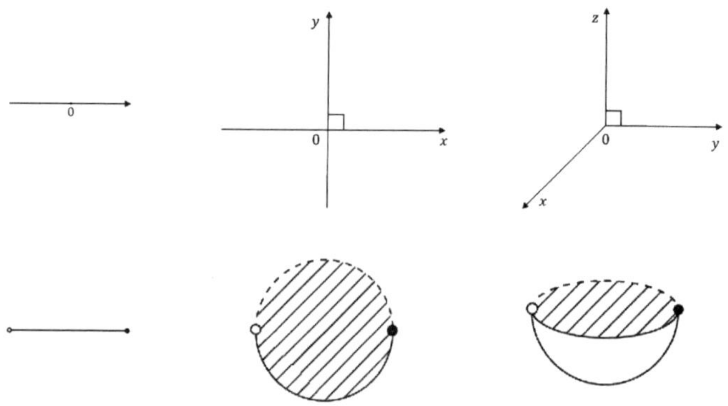

$A_1 = \mathbb{R} \cap A_2$ $\quad$ $A_2 = \mathbb{R}^2 \cap A_3$ $\quad$ $A_3 \subset \mathbb{R}^3$

We can find the following subsets :

$i(A_1),\ e(A_1),\ b(A_1)$ $\quad$ $i(A_2),\ e(A_2),\ b(A_2)$ $\quad$ $i(A_3),\ e(A_3),\ b(A_3)$

$\overline{A_1},\ A_1'$ $\quad$ $\overline{A_2},\ A_2'$ $\quad$ $\overline{A_3},\ A_3'$

$i(A_1),\ e(A_1),\ b(A_1)$ $\quad$ $i(A_2),\ e(A_2),\ b(A_2)$

$\overline{A_1},\ A_1'$ $\quad$ $\overline{A_2},\ A_2'$

$i(A_1),\ e(A_1),\ b(A_1)$

$\overline{A_1},\ A_1'$

2.3. Exercises

1. Let $(X, \mathcal{T})$ be a topological space, and $A \subset X$. Then show

(1) $b(A) = \bar{A} \cap \overline{(X - A)}$

(2) $\bar{A} = i(A) \cup b(A)$

(3) $b(A) = \emptyset$ iff A is open and closed.

(4) If A is open, is it true $A = i(\bar{A})$?

2. Let A_1 and A_2 be dense in X. Then prove that if A_1 and A_2 are open, then $A_1 \cap A_2$ is dense in X.

3. In Example 2.3.4, find the sets $i(A_j)$, $e(A_j)$, $b(A_j)$, A'_j and $\overline{A_j}$ in $\mathbb{R}^j$ for $j = 1, 2, 3$, respectively.

2.4 Basis and Subbasis

Definition 2.4.1. Let $(X, \mathcal{T})$ be a topological space.

(1) A subfamily $\mathfrak{B} \subset \mathcal{T}$ is a **basis** for $\mathcal{T}$ if for each open set $U \in \mathcal{T}$ there is $\mathfrak{B}_1 \subset \mathfrak{B}$ such that $U = \bigcup_{B_\alpha \in \mathfrak{B}_1} B_\alpha$. An element $B \in \mathfrak{B}$ is called a **basic open set**.

(2) A subfamily $\mathcal{S} \subset \mathcal{T}$ is a **subbasis** for $\mathcal{T}$ if the family of finite intersections of $\mathcal{S}$ forms a basis for $\mathcal{T}$. That is, every open set is a union of finite intersections of elements of S. An element of S is called a **subbasic open set**.

(3) For a point $x \in X$ and $\mathfrak{N}_x$ the neighborhood system of x, a subfamily $\mathfrak{B}_x \subset \mathfrak{N}_x$ is a **local basis** of $\mathfrak{N}_x$ if for each $U \subset \mathfrak{N}_x$ there is $V \in \mathfrak{B}_x$ such that $V \subset U$.

Theorem 2.4.2. *Let X be a set and $\mathfrak{B}$ be a family of subsets of X. Then $\mathfrak{B}$ is a basis for a topology $\mathcal{T}$ of X if and only if*

(1) $X = \bigcup\limits_{B \in \mathfrak{B}} B$

(2) *If B_1, $B_2 \in \mathfrak{B}$ and $x \in B_1 \bigcap B_2$, then there is $B \in \mathfrak{B}$ such that $x \in B \subset B_1 \cap B_2$.*

Proof. ($\Rightarrow$) Suppose that $\mathfrak{B}$ is a basis for a topology $\mathcal{T}$ of X. Since $X \in \mathcal{T}$, $X = \bigcup_{B \in \mathfrak{B}} B$. If B_1, $B_2 \in \mathfrak{B}$, and $x \in B_1 \cap B_2$. Since $\mathfrak{B} \subset \mathcal{T}$, $B_1 \cap B_2 \in \mathcal{T}$. There is a subfamily $\mathfrak{B}_1 \subset \mathfrak{B}$ such that $B_1 \cap B_2 = \bigcup_{B_\alpha \in \mathfrak{B}_1} B_\alpha$.
Thus there is $B_\alpha \in \mathfrak{B}_1$ such that $x \in B_\alpha \subset B_1 \bigcap B_2$.

($\Leftarrow$) Define

$$\mathcal{T} := \Big\{ \bigcup_{B \in \mathfrak{B}_1} B \mid \mathfrak{B}_1 \subset \mathfrak{B} \Big\}.$$

Then $\mathcal{T}$ is a topology of X and $\mathfrak{B}$ is a basis for the topology $\mathcal{T}$ of X. Indeed, show $\mathfrak{B}$ is a basis.

(i) Take $\mathfrak{B}_1 = \mathfrak{B}$, then $X = \bigcup_{B \in \mathfrak{B}_1} B \in \mathcal{T}$ by (2). Take $\mathfrak{B}_1 = \emptyset \subset \mathfrak{B}$, then $\emptyset = \bigcup_{B \in \mathfrak{B}_1} B \in \mathcal{T}$.

(ii) Suppose $\left\{ \bigcup_{B \in \mathfrak{B}_\alpha} B \right\}_{\alpha \in I}$ be a family of $\mathcal{T}$. Then for each $\alpha \in I$, $\mathfrak{B}_\alpha \subset \mathfrak{B}$ and $\mathfrak{B}_1 = \bigcup_{\alpha \in I} \mathfrak{B}_\alpha \subset \mathfrak{B}$. Thus

$$\bigcup_{B \in \mathfrak{B}_1} B = \bigcup_{\alpha \in I} \Big(\bigcup_{B \in \mathfrak{B}_\alpha} B \Big) \in \mathcal{T}.$$

(iii) Let $\left\{ \bigcup_{B \in \mathfrak{B}_i} B \mid i = 1, 2, \cdots, n \right\}$ be a finite family of $\mathcal{T}$. Then

$$\bigcap_{i=1}^{n} \Big(\bigcup_{B \in \mathfrak{B}_i} \Big) = \bigcup_{B \in \mathfrak{B}'} B$$

for some $\mathfrak{B}' \subset \mathfrak{B}$ by (2). Thus $\bigcap_{i=1}^{n} \left(\bigcup_{B \in \mathfrak{B}_i} B \right) \in \mathcal{T}$.

□

Theorem 2.4.3. *Let $\mathcal{S}$ be a family of subsets of X, and whose union X. Then there is a unique topology $\mathcal{T}$ of X such that $\mathcal{S}$ is a subbasis for $\mathcal{T}$.*

Proof. Let $\mathfrak{B} := \left\{ \bigcap_{S \in \mathcal{S}'} S \mid \mathcal{S}' \text{ is a finite subset of } \mathcal{S} \right\}$.
Show that $\mathfrak{B}$ satisfies Theorem 2.4.2 (1) and (2).

(1) By assumption

$$\bigcup_{S \in \mathcal{S}} S = X \in \mathfrak{B}.$$

(2) Let $B_1 = \bigcap_{S \in \mathcal{S}_1} S$, $B_2 = \bigcap_{S \in \mathcal{S}_2} S \in \mathfrak{B}$. Then $\mathcal{S}' = \mathcal{S}_1 \cup \mathcal{S}_2$ is finite, and

$$B_1 \bigcap B_2 = \Big(\bigcap_{S \in \mathcal{S}_1} S \Big) \bigcap \Big(\bigcap_{S \in \mathcal{S}_2} S \Big) = \bigcap_{S \in \mathcal{S}'} S \in \mathfrak{B}.$$

Therefore $\mathfrak{B}$ is a basis for a topology $\mathcal{T}$ on X, also every element of $\mathcal{T}$ is a union of finite intersections of elements of $\mathcal{S}$.

□

Example 2.4.4. (1) For $(\mathbb{R}, \mathcal{T}_{d_1})$,

$$\mathfrak{B} = \{(a,b) \mid a, b \in \mathbb{Q}\} \text{ is a basis for } \mathcal{T}_{d_1}.$$
$$\mathcal{S} = \{(a,\infty), (-\infty, b) \mid a, b \in \mathbb{Q}\} \text{ is a basis for } \mathcal{T}_{d_1}.$$

(2) For $(\mathbb{R}^2, \mathcal{T}_{d_2})$,

$$\mathfrak{B}_1 = \{B(p,\varepsilon) \mid p \in \mathbb{R}^2, \varepsilon > 0\} \text{ is a basis for } \mathcal{T}_{d_2}.$$
$$\mathfrak{B}_2 = \{B(q,r) \mid q \in \text{rational points}, r > 0, \text{ rational}\}$$
$$\text{is a basis for } \mathcal{T}_{d_2}.$$
$$\mathfrak{B}_3 = \{(a,b) \times (c,d) \mid a, b, c, d \in \mathbb{Q}\} \text{ is a basis for } \mathcal{T}_{d_2}.$$
$$\mathcal{S} = \{(x,y) \mid y > c\}_{c\in\mathbb{Q}} \bigcup \{(x,y) \mid d > y\}_{d\in\mathbb{Q}}$$
$$\bigcup \{(x,y) \mid x > a\}_{a\in\mathbb{Q}} \bigcup \{(x,y) \mid b > x\}_{b\in\mathbb{Q}}$$
$$\text{is a subbasis for } \mathcal{T}_{d_2}.$$

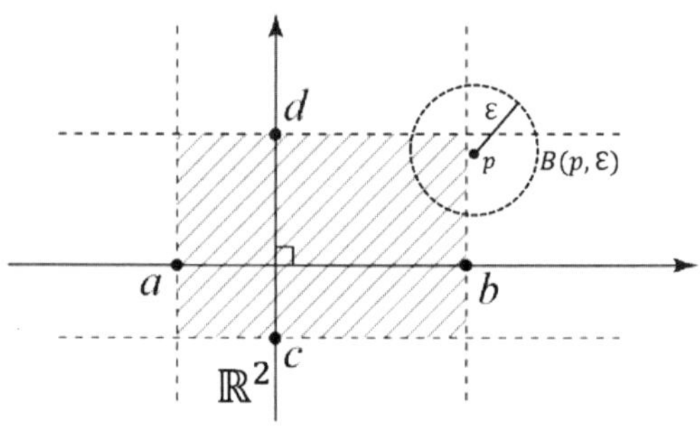

2.4. Exercises

1. Let $\mathfrak{B}$ be a basis for a topological space $(X, \mathcal{T})$ and $A \subset \mathcal{T}$. Show that $\mathfrak{B}_A = \{A \cap U \mid U \in \mathfrak{B}\}$ is a basis for the subspace $(A, \mathcal{T}_A)$.

2. Show that if $\mathcal{S}$ is a subbasis for a topology $\mathcal{T}$ on X, then $\mathcal{S} - \{X, \emptyset\}$ is also a subbasis for $\mathcal{T}$.

3. Let $X = \{x, y, z, w\}$ be a set. Find the topology $\mathcal{T}$ on X generated by

$$\{\{x\}, \{x, y, z\}, \{z, w\}\}.$$

4. Let $(X, \mathcal{T})$ be a topological space with a basis $\mathfrak{B}$ and a subbasis $\mathcal{S}$ for the topology $\mathcal{T}$, and $p \in X$ $\mathfrak{N}_p$ the nbd system of p. Then show that

(1) $\mathfrak{B}_p = \{B \in \mathfrak{B} \,|\, p \in B\}$ is a local basis at p.

(2) $\mathcal{S}_p = \{S \in \mathcal{S} \,|\, p \in S\}$ need not be a local basis at p.

(3) A sequence (a_n) in X converges to a iff every $S \in \mathcal{S}_p$ contains all except a finite number of terms of the sequence.

5. Prove Example 2.4.4.

Chapter 3

Continuous Maps

3.1 Continuous Maps

Recall a real valued map $f \ \mathbb{R} \longrightarrow \mathbb{R}$ is continuous at $x = a$

$$\begin{aligned} &\text{iff} \quad \lim_{x \to a} f(x) = f(a) \\ &\text{iff} \quad \text{for any } \varepsilon > 0, \text{ there is } \delta > 0 \text{ such that} \\ &\qquad\qquad |f(x) - f(a)| < \varepsilon \text{ if } |x - a| < \delta. \end{aligned}$$

Definition 3.1.1. Let $f : (X, \mathcal{T}_X) \longrightarrow (Y, \mathcal{T}_Y)$ be a map.

(1) f is **continuous**, if $V \in \mathcal{T}_Y$, then $f^{-1}(V) \in \mathcal{T}_X$.

(2) f is **continuous at** $x \in X$ if V is a nbd of $f(x)$, then there is a neighborhood U of x such that $f(U) \subset V$.

Theorem 3.1.2. *For a map* $f \ (X, \mathcal{T}_X) \longrightarrow (Y, \mathcal{T}_Y)$, *the followings are equivalent :*

(1) *f is continuous.*

(2) *If B is closed in Y, then $f^{-1}(B)$ is closed in X.*

(3) *If B is a basic open in Y, then $f^{-1}(B)$ is open in X.*

(4) *If S is subbasic open in Y, then $f^{-1}(S)$ is open in X.*

(5) *For any subset $A \subset X$, $f(\bar{A}) \subset \overline{f(A)}$.*

(6) *f is continuous at all points in X.*

(7) *For any subset $B \subset Y$, $f^{-1}(iB) \subset if^{-1}(B)$.*

Proof. (1) $\sim$ (4) are clear, because

$$f^{-1}(Y - V) = X - f^{-1}(V),$$

$$f^{-1}(\bigcup_{\alpha} V_\alpha) = \bigcup_{\alpha} f(V_\alpha),$$

and

$$f^{-1}(\bigcap_{\alpha} V_\alpha) = \bigcap_{\alpha} f(V_\alpha).$$

(5) Since $f(A) \subset \overline{f(A)}$, $A \subset f^{-1}(f(A)) \subset f^{-1}(\overline{f(A)})$.
Since f is continuous, $f^{-1}(\overline{f(A)})$ is closed and $\bar{A} \subset f^{-1}(\overline{f(A)})$.
Thus $f(\bar{A}) \subset ff^{-1}(\overline{f(A)}) \subset \overline{f(A)}$.

Conversely, suppose that B is closed in Y and $A = f^{-1}(B)$. Then

$$f(\bar{A}) \subset \overline{f(A)} = \overline{ff^{-1}(B)} \subset \bar{B} = B,$$

and

$$\bar{A} \subset f^{-1}(f(\bar{A})) \subset f^{-1}(B) = A.$$

Thus $A = f^{-1}(B)$ is closed in X.

(6) Suppose that f is continuous and $x \in X$. If V is a nbd of $f(x)$ in Y, then $f^{-1}(V)$ is open and a nbd of x.
Thus $f(f^{-1}(V)) \subset V$ and f is continuous at x.

Conversely, suppose that f is continuous at all points $x \in X$. If V is an open set in Y, and $x \in f^{-1}(V)$, then there is $U(x)$ a nbd of x such that $f(U(x)) \subset V$. So,

$$U(x) \subset f^{-1}(f(U(x)) \subset f^{-1}(V)$$

and $f^{-1}(V) = \bigcup_{x \in f^{-1}(V)} U(x)$ is open in X.
Thus f is continuous.

(1) $\Leftrightarrow$ (7)

($\Rightarrow$) If $f : X \to Y$ is continuous, and $B \subset Y$, then $f^{-1}(iB)$ is open and $f^{-1}(iB) \subset f^{-1}(B)$. Since $if^{-1}(B)$ is the largest open subset in $f^{-1}(B)$, $f^{-1}(iB) \subset if^{-1}(B)$.

($\Leftarrow$) If V is open in Y, then $iV = V$ and

$$f^{-1}(V) = f^{-1}(iV) \subset if^{-1}(V)$$

by hypothesis.
Thus $f^{-1}(V) = if^{-1}(V)$ is open in X. And so f is continuous. □

Definition 3.1.3. Let $\mathbb{N}$ be the set of natural numbers.

(1) A map $x : \mathbb{N} \longrightarrow X$ is called a **sequence of points** in X.
Denoted it by $x(n) = x_n$ and $x = < x_n >$.

(2) For natural numbers $n_1 < n_2 < \cdots$, the sequence $< x_{n_i} >$ is called a **subsequence** of $< x_n >$.

(3) A sequence $< x_n >$ **converges to** a point x, if for any nbd U of x there is $n_0 \in \mathbb{N}$ such that $x_n \in U$ for all $n > n_0$. In this case we denoted it by $\lim_{n\to\infty} x_n = x$, and the x is called the **limit of** $< x_n >$.

(4) A map $f : X \longrightarrow Y$ is **sequentially continuous at** x iff

$$\lim_{n\to\infty} f(x_n) = f(x),$$

when $\lim_{n\to\infty} x_n = x$.

Theorem 3.1.4. *If a map $f : X \longrightarrow Y$ is continuous at $x \in X$, then f is sequentially continuous at x.*

Proof. Suppose that a sequence $< x_n >$ converges to x in X. Let V be a nbd of $f(x)$. There is a nbd U of x such that $f(U) \subset V$. There is $n_0 \in \mathbb{Z}$ such that $x_n \in U$ for all $n > n_0$. Therefore $f(x_n) \in f(U) \subset V$ for all $n > n_0$, and $\lim_{n\to\infty} f(x_n) = f(x)$.
Thus f is sequentially continuous at x. □

Example 3.1.5. Let $\mathbb{R}$ be the set of real numbers and

$$\mathcal{T} = \{\emptyset\} \cup \{\mathbb{R} - A \,|\, A : \text{ countable}\}.$$

Then $\mathcal{T}$ is a topology on $\mathbb{R}$. A sequence $< a_n >$ in $(\mathbb{R}, \mathcal{T})$ converges to a iff there is n_0 such that

$$< a_n >= (a_1, \cdots, a_{n_0}, a, a, a, \cdots).$$

$$(\text{i.e., } a_n = a, \ \forall n > n_0 \text{ for some } n_0)$$

Proof. ($\Leftarrow$) It is clear.
($\Rightarrow$) Let $\lim_{n\to\infty} a_n = a$.
Suppose that there is no n_0 such that $a_n = a$, for all $n > n_0$.
Then for any n_0, there is n such that $a_n \neq a$, $n > n_0$.
Let

$$A = \{a_{n'} \in < a_n > \,|\, a_{n'} \neq a\}.$$

Then A countable and for any n_0 there is $n > n_0$ such that $a_n \neq a$, $a_n \in A$.

The set $\mathbb{R} - A$ is open and contains a, i.e., $\mathbb{R} - A$ is a nbd of a but there is n_0 such that $a_n \in \mathbb{R} - A$ for all $n > n_0$. Thus

$$\lim_{n\to\infty} a_n \neq a,$$

which makes a contradiction.

Therefore the map $f : (\mathbb{R}, \mathcal{T}) \longrightarrow (\mathbb{R}, d)$ given by $f(x) = x$ for all $x \in \mathbb{R}$, is sequentially continuous but not continuous. □

Theorem 3.1.6. *Let* $X = \bigcup_{i=1}^{n} F_i$. F_i *are closed,* $i = 1, 2, \cdots, n$. *A map* $f : X \longrightarrow Y$ *is continuous iff the restrictions* $f_{F_i} : F_i \longrightarrow Y$ *are continuous for* $i = 1, \cdots, n$.

Proof. ($\Rightarrow$) Let G be a closed set in Y. For each $i = 1, 2, \cdots, n$,

$$(f_{F_i})^{-1}(G) = F_i \bigcap f^{-1}(G)$$

is closed in F_i, since $f^{-1}(G)$ is closed in X. Thus f_{F_i} is continuous for each $i = 1, \cdots, n$.

($\Leftarrow$) Suppose that $f_{F_i} : F_i \longrightarrow Y$ is continuous for each $i = 1, \cdots, n$. If G is closed in Y, then

$$f_{F_i}^{-1}(G) = F_i \bigcap f^{-1}(G)$$

is closed in F_i and also closed in X.

The union

$$\bigcup_{i=1}^{n} (f_{F_i}^{-1}(G)) = \bigcup_{i=1}^{n} (F_i \bigcap f^{-1}(G))$$
$$= (\bigcup_{i=1}^{n} F_i) \cap f^{-1}(G) = X \bigcap f^{-1}(G) = f^{-1}(G)$$

is closed in X. Thus f is continuous on X. □

Example 3.1.7. Let $f : (\mathbb{R}, d) \longrightarrow (\mathbb{R}, d)$ be defined by

$$f(x) = \begin{cases} 1, & x \geq 0 \\ -1, & x < 0 \end{cases}$$

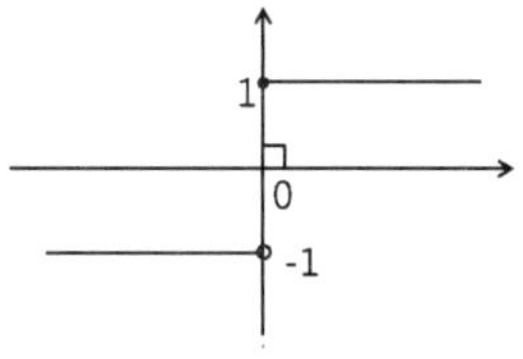

$F_0 = [0, \infty)$, $F_n = (-\infty, -\frac{1}{n}]$, then $\mathbb{R} = \bigcup_{n=0}^{\infty} F_n$.

The restriction $f_{F_i} : F_i \longrightarrow \mathbb{R}$ are continuous, $i = 0, 1, \cdots$, since all maps are constant. However, $f : \mathbb{R} \longrightarrow \mathbb{R}$ is not continuous at $x = 0$, since $\lim_{x \to 0} f(x) \neq f(0) = 1$.

3.1. Exercises

1. Let the maps $f : X \longrightarrow Y$ and $g : Y \longrightarrow Z$ be continuous. Prove the composition map $g \circ f : X \longrightarrow Z$ is continuous.

2. Let $f : (X, \mathcal{T}_X) \longrightarrow (Y, \mathcal{T}_Y$ be continuous and $A \subset X$. Prove the restriction $f_A : (A, \mathcal{T}_A) \longrightarrow (Y, \mathcal{T}_Y)$ is continuous.

3. Prove that a map $f : X \longrightarrow Y$ is continuous iff

$$f^{-1}\big(i(A)\big) \subset i\big[f^{-1}(A)\big]$$

for every $A \subset Y$.

4. Let X and Y be topological spaces, $X = X_1 \cup X_2$, and $f_1 : X_1 \to Y$, $f_2 : X_2 \to Y$ be continuous and $f_1 = f_2$ on $X_1 \cap X_2$. Let $f : X \to Y$ be the map defined by

$$f(x) = \begin{cases} f_1(x), & x \in X_1 \\ f_2(x), & x \in X_2 \end{cases}$$

Prove the followings.

(1) f need not be continuous.

(2) f is continuous if X_1 and X_2 are open.

(3) f is continuous if X_1 and X_2 are closed.

3.2 Topological Properties

Definition 3.2.1. For a map $f : X \to Y$ between topological spaces,

(1) f is an **open** map if $f(U)$ is open in Y when U is open in X.

(2) f is a **closed** map if $f(F)$ is closed in Y when F is closed in X.

(3) f is a **homeomorphism** if f is bijective and continuous and its inverse f^{-1} is also continuous.

(4) A property which is invariant under homeomorphisms is called a **topological property** or a **topological invariant**.

Remark 3.2.2. (1) If $f : X \longrightarrow Y$ is a homeomorphism, then $f^{-1} : Y \longrightarrow X$ is also a homeomorphism.

(2) If f is a homeomorphism, then f is open and closed.

Example 3.2.3. (1) The map $f : (-1, 1) \longrightarrow \mathbb{R}$ given by

$$f(x) = \tan \frac{\pi}{2} x$$

is a homeomorphism.

(2) The map $f : (\mathbb{R}, \mathfrak{D}) \longrightarrow (\mathbb{R}, d)$, $f(x) = x$, is continuous and bijective but not a homeomorphism.

(3) In Example 3.1.5, the map $f : (\mathbb{R}, \mathcal{T}) \longrightarrow (\mathbb{R}, d)$ is bijective but f and f^{-1} are not continuous.

(4) Openness and closedness are topological properties, but distance and Cauchy sequence are not.

Theorem 3.2.4. *For a map $f : X \longrightarrow Y$ between topological spaces,*

(1) *f is open iff for any $A \subset X$, $f\big(i(A)\big) \subset i\big(f(A)\big)$.*

(2) *f is closed iff for any $A \subset X$, $\overline{f(A)} \subset f(\bar{A})$.*

(3) *If f is continuous, bijective and open (or closed),*

then f is a homeomorphism.

(4) *If f is a homeomorphism, $f(A) = B$, then $f_A : A \longrightarrow B$ is also a homeomorphism.*

Proof. (1) ($\Rightarrow$) Since $i(A) \subset A$, $f(i(A)) \subset f(A)$.

Since f is open, $f(i(A))$ is open and $f(i(A)) \subset i(f(A))$.

($\Leftarrow$) If A is open, then $A = i(A)$. So,

$$f(A) = f(i(A)) \subset i(f(A)) \subset f(A).$$

Thus $f(A) = if(A)$ and $f(A)$ is open.

(2) The proof is similar to (1).

(3) Since $f : X \longrightarrow Y$ is continuous, bijective and open, there is the inverse map $f^{-1} : Y \longrightarrow X$. If U is open in X, then $(f^{-1})^{-1}(U) = f(U)$ is open and f^{-1} is continuous.
Therefore f is a homeomorphism.

(4) The restriction $f_A : A \longrightarrow B = f(A)$ is continuous, bijective and open, since if U is open in X, then

$$\begin{aligned} f(A \cap U) &= f(A) \cap f(U) \\ &= B \cap f(U) = f_A(U). \end{aligned}$$

Thus f_A is a homeomorphism.

□

Definition 3.2.5. For a map $f : X \longrightarrow Y$,

(1) Let X be a set and $(Y, \mathcal{T}_Y)$ a topological space. Then

$$\mathcal{T}_X := \{f^{-1}(V) \mid V \in \mathcal{T}_Y\}$$

is called the **induced topology** on X with respect to f.

(2) Let $(X, \mathcal{T}_X)$ be a topological space and Y a set. Then

$$\mathcal{T}_Y := \{V \subset Y \mid f^{-1}(V) \in \mathcal{T}_X\}$$

is called the **identification topology** on Y with respect to f.

Remark 3.2.6. (1) In Definition 3.2.5 (1),

(i) $\mathcal{T}_X$ is a topology on X.

(ii) f is continuous.

(iii) $\mathcal{T}_X$ is the smallest topology on X such that f is continuous.

(2) In Definition 3.2.5 (2),

(i) $\mathcal{T}_Y$ is a topology on Y.

(ii) f is continuous.

(iii) $\mathcal{T}_Y$ is the largest topology on Y such that f is continuous.

Theorem 3.2.7. *For maps* $(X, \mathcal{T}_X) \xrightarrow{f} (Y, \mathcal{T}_Y) \xrightarrow{g} (Z, \mathcal{T}_Z)$,

(1) *Let* $\mathcal{T}_Y$ *be the induced topology with respect to* g. *Then* f *is continuous iff* $g \circ f$ *is continuous.*

(2) *If* f *is continuous, surjective and open, then* $\mathcal{T}_Y$ *is the identification topology on* Y *with respect to* f.

(3) *Let* f *be surjective and* $\mathcal{T}_Y$ *the identification topology on* Y *with respect to* f. *Then* g *is continuous iff* $g \circ f$ *is continuous.*

Proof. (1) By Remark 3.2.6 (1), g is continuous.

($\Rightarrow$) If f is continuous, then the composite $g \circ f$ is continuous.

($\Leftarrow$) If $V \in \mathcal{T}_Y$, then there is $W \in \mathcal{T}_Z$ such that $V = g{-}1(W)$. Thus

$$f^{-1}(V) = f^{-1}(g^{-1}(W)) = (g \circ f)^{-1}(W)$$

is open.

(2) Suppose that $f : (X, \mathcal{T}_X) \longrightarrow (Y, \mathcal{T}_Y)$ is continuous, open and surjective. Show $V \in \mathcal{T}_Y$ iff $f^{-1}(V) \in \mathcal{T}_X$.

($\Leftarrow$) Since f is open, $f(f^{-1}(V)) \in \mathcal{T}_Y$.
Since f is onto, $V = f(f^{-1}(V))$.
Thus $V \in \mathcal{T}_Y$.

(3) ($\Rightarrow$) Since $\mathcal{T}_Y$ is an identification topology on Y with respect to f, f is continuous, so $g \circ f$ is continuous.

($\Leftarrow$) Suppose that $g \circ f$ is continuous. If W is open in Z, then $(g \circ f)^{-1}(W) = f^{-1}(g^{-1}(W))$ is open in X. Since $\mathcal{T}_Y$ is the identification topology on Y with respect to f, $g^{-1}(W)$ is open in Y.

□

3.2. Exercises

1. Let $f : X \longrightarrow Y$ and $g : Y \longrightarrow Z$ be maps between topological spaces. Prove the followings:

(1) If f, g are open (closed), then $g \circ f$ is also open (closed).

(2) If $g \circ f$ is open and f is continuous and surjective, then g is open.

(3) If $g \circ f$ is open and g is continuous and injective, then f is open.

2. Let a map $f : (X, \mathcal{T}_X) \longrightarrow (Y, \mathcal{T}_Y)$ be surjective and open. If $\mathfrak{B}$ is a basis for $\mathcal{T}_X$, then show that $\{f(U) \mid U \in \mathfrak{B}\}$ is a basis for $\mathcal{T}_Y$.

3. Show that limit point, interior, boundary, convergence, and density are topological properties.

4. Show that distance, angle, Cauchy sequence, and area are not topological properties.

Chapter 4

Product Spaces and Quotient Spaces

4.1 Product Spaces

(1) Recall, let X be a set and $(Y, \mathcal{T}_Y)$ a topological space, and let $f : X \longrightarrow Y$ be a map. Then

$$\mathcal{T}_X = \{f^{-1}(V) \mid V \in \mathcal{T}_Y\}$$

is called the **induced topology** on X with respect to f.

(2) Let $(X_1, \mathcal{T}_{X_1})$, $(X_2, \mathcal{T}_{X_2})$ be topological spaces. Let $X_1 \times X_2 \xrightarrow{p_i} X_i$ be projections, $i = 1, 2$. Then

$$\mathcal{S}_{X_1 \times X_2} = \{p_i^{-1}(U_i \mid U_i \in \mathcal{T}_{X_i}, i = 1, 2\}$$

is a **subbasis for a topology** on $X_1 \times X_2$.

(3) Let $\{X_i \mid i \in I\}$ be a family of sets. The **product set**

$$\prod_{i \in I} X_i = \left\{x : I \longrightarrow \bigcup_{i \in I} X_i \mid x(i) \in X_i, i \in I\right\} \xrightarrow{p_i} X_i$$

where p_i is a **projection** defined by $p_i(x) = x(i) \in X_i$.

Definition 4.1.1. Let $\mathfrak{F} = \{(X_i, \mathcal{T}_i) \mid i \in I\}$ be a family of spaces. Let a topology $\mathcal{T}$ on the product $\prod_{i\in I} X_i$ be defined by the topology with a subbasis $\mathcal{S} = \{p_i^{-1}(U_j) \mid i \in I,\ U_j \in \mathcal{T}_i\}$.
Then $(\prod_{i\in I} X_i, \mathcal{T})$ is called the **product space** of $\mathfrak{F}$,
and $\mathcal{T}$ is called the **product topology** on $\prod_{i\in I} X_i$.

Remark 4.1.2. For the product space $(\prod_{i\in I} X_i, \mathcal{T})$,

(1) The **subbasic open set** is of the form :

$$p_i^{-1}(U_i) = \prod_{i\in I} V_i \in \mathcal{S},$$

where $V_j = \begin{cases} U_i, & i = j \\ X_j, & i \neq j. \end{cases}$

(2) The basis $\mathfrak{B}$ of $\mathcal{T}$ is

$$\mathfrak{B} := \Big\{ \bigcap_{i\in I_1} p_i^{-1}(U_i) \mid I_1 \text{ is a finite subset of } I,\ U_i \in \mathcal{T}_i \Big\}.$$

A **basic open set** is of a form $\bigcap_{i\in I_1} p_i^{-1}(U_i) = \prod_{i\in I} V_i$, where

$$V_j = \begin{cases} \bigcap_{j=k} U_k, & j \in I_1 \\ X_j, & j \in I - I_1. \end{cases}$$

Example 4.1.3. (1) Let $(\mathbb{R}, \mathfrak{U})$ be the usual real line. Then

$\mathfrak{B} = \{(a, b) \mid a, b \in \mathbb{R}\}$ is a basis for $\mathfrak{U}$ and

$\mathcal{S} = \{(a, \infty), (-\infty, b) \mid a, b \in \mathbb{R}\}$ is a subbasis for $\mathfrak{U}$.

(2) Let $(\mathbb{R}^2, \mathcal{T}) \xrightarrow{p_i} \mathbb{R}$ be a projection, $i = 1, 2$. Then

$\mathfrak{B} = \{(a,b) \times (c,d) \mid a,b,c,d \in \mathbb{R}\}$ is a basis, and

$\mathcal{S} = \{(a,b) \times \mathbb{R}, \mathbb{R} \times (c,b) \mid a,b,c,d \in \mathbb{R}\}$ is a subbasis.

$p_1^{-1}\big((a,b)\big) = (a,b) \times \mathbb{R}, \quad p_2^{-1}\big((c,d)\big) = \mathbb{R} \times (c,d),$

$p_1^{-1}\big((a,b)\big) \bigcap p_2^{-1}\big((c,d)\big) = (a,b) \times (c,d).$

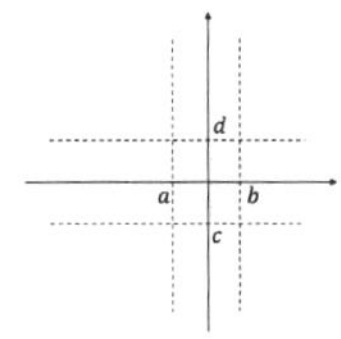

(3) Let $I = [0,1]$ and $X_i = \mathbb{R}$ for all $i \in I$. Then the product space $\prod_{i \in I} X_i = \{x : I \to \mathbb{R}\}$ has a subbasis

$$\mathcal{S} = \{p_i^{-1}(U) \mid U : \text{ open in } \mathbb{R}, \quad i \in I\}.$$

If $(a_i, b_i) \subset \mathbb{R}$, $i \in I$, then $p_i^{-1}\big((a_i, b_i)\big)$ is open, but $\prod_{i \in I}(a_i, b_i)$ is not open.

Theorem 4.1.4. *For a family of topological spaces $\{(X_i, \mathcal{T}_i) \mid i \in I\}$ and the product space $\prod_{i \in I} X_i$,*

(1) *the projections $p_i : \prod_{i \in I} X_i \longrightarrow X_i$ are open.*

(2) *a map $f : (Y, \mathcal{T}_Y) \longrightarrow \prod_{i \in I} X_i$ is continuous iff $p_i \circ f$ is continuous for each $i \in I$.*

Proof. (1) The projection $p_i : \prod\limits_{i \in I} X_i \longrightarrow X_i$ is onto. For a subbasic open $p_i^{-1}(U_i)$, the image is

$$p_j\big(p_i^{-1}(U_i)\big) = \begin{cases} U_i, & i = j \\ X_j, & i \neq j. \end{cases}$$

For any open $U \subset \prod_{i \in I} X_i$, $p_j(U)$ is open.

(2) $(\Rightarrow)$ If f is continuous, then $p_i \circ f$ is continuous for each $i \in I$.

$(\Leftarrow)$ Let $p_i^{-1}(U_i)$ be a subbasic open. Then

$$f^{-1}\big(p_i^{-1}(U_i)\big) = (p_i \circ f)^{-1}(U_i)$$

is open. So, f is continuous.

□

Theorem 4.1.5. *Let $\{f_i : X_i \longrightarrow Y_i \mid i \in I\}$ be a family of maps between topological spaces. Then the product f given by*

$$\begin{array}{ccc} \prod\limits_{i\in I} X_i & \xrightarrow{f := \prod\limits_{i\in I} f_i} & \prod\limits_{i\in I} Y_i \\ \downarrow{\scriptstyle p_i} & & \downarrow{\scriptstyle q_i} \\ X_i & \xrightarrow{f_i} & Y_i \end{array}$$

is continuous iff f_i is continuous for each $i \in I$, where p_i and q_i are projections, $\prod X_i$ and $\prod Y_i$ have product topologies, and $f(x)(i) := f_i(x_i)$ for each $i \in I$, $x \in \prod_{i\in I} X_i$.

Proof. Since for each $i \in I$, $x \in \prod_{i\in I} X_i$,

$$(q_i \circ f)(x) = q_i(f(x)) = f(x)(i) = f_i(x_i) = (f_i \circ p_i)(x).$$

Thus $q_i \circ f = f_i \circ p_i$ for each $i \in I$. Since the projections p_i, q_i are continuous for each $i \in I$,

$$\begin{aligned} f \text{ is continuous } &\text{iff } q_i \circ f \text{ is continuous, } \forall i \in I. \\ &\text{iff } f_i \circ p_i \text{ is continuous, } \forall i \in I. \\ &\text{iff } f_i \text{ is continuous, } \forall i \in I. \end{aligned}$$

□

4.1. Exercises

1. Let $\{X_\alpha \mid \alpha \in I\}$ be a family of topological spaces and $A_\alpha \subset X_\alpha$ for each $\alpha \in I$. Then prove the followings:

(1) If A_α is closed in X_α for each $\alpha \in I$,
then $\prod\limits_{i\in I} A_\alpha$ is closed in $\prod\limits_{i\in I} X_\alpha$.

(2) If A_α is open in X_α for each $\alpha \in I$,
then $\prod_{i \in I} A_\alpha$ may not be open in $\prod_{i \in I} X_\alpha$.

(3) $\overline{\prod_{i \in I} A_\alpha} = \prod_{i \in I} \overline{A_\alpha}$.

2. Show that X is Hausdorff (i.e., distinct two points of X have disjoint nbds in X) iff the diagonal

$$\Delta = \{(x, x) \mid x \in X\}$$

is closed in $X \times X$.

4.2 Quotient Spaces

Definition 4.2.1. Let $(X, \mathcal{T})$ be a topological space. Let $\sim$ be an equivalence relation on X, and

$$X/\sim = \{[x] \mid x \in X\}$$

be the set of equivalence classes. The map $p : X \longrightarrow X/\sim$ given by $p(x) = [x]$ defines the **identification topology** $\mathcal{T}'$ on $X/\sim$, that is, $V \subset X/\sim$ is open iff $p^{-1}(V)$ is open in X.
Then we call $(X/\sim, \mathcal{T}')$ the **quotient space**, p the **quotient map**, and $\mathcal{T}'$ the **quotient topology**.

Remark 4.2.2. In the definition 4.2.1, the equivalence class

$$[x] = \{y \in X \mid x \sim y\} \subset X$$

is a subset in X, but $[x] \in X/\sim$ is an element in $X/\sim$.

Example 4.2.3. On $\mathbb{R}$, define $x \sim y$ iff $x - y \in \mathbb{Z}$. Then $\sim$ is an equivalence relation on $\mathbb{R}$.

Let $p : \mathbb{R} \longrightarrow \mathbb{R}/\sim$ be the projection. Then the following diagram is commutative.

$$\begin{array}{ccc} \mathbb{R} & \xrightarrow{\ p\ } & \mathbb{R}/\sim \\ & \searrow^{e^{2\pi i\,\cdot}} & \downarrow \pi \\ & & S^1 \end{array}$$

Indeed, for each $t \in \mathbb{R}$ $\pi(p(t)) = \pi\big([t]\big) = e^{2\pi i t}$. Since $[t] = \{t+n \mid n \in \mathbb{Z}\}$, π is well-defined. If we define an operation on $\mathbb{R}/\sim$ by $[t_1]+[t_2] = [t_1+t_2]$, then $\mathbb{R}/\sim$ is a group. Moreover, the map $\pi : \mathbb{R}/\sim \longrightarrow S^1$ is a homeomorphism and an isomorphism.

Example 4.2.4. On $S^2 = \{x \in \mathbb{R}^2 \,|\, |x| = 1\}$, define an equivalence relation $x \sim y$ iff $y = \pm x$. The projection

$$p : S^2 \longrightarrow S^2/\sim\, = \mathbb{RP}^2$$

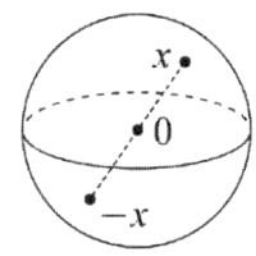

is given by

$$p(x) = [x] \in \mathbb{RP}^2 \quad \text{and} \quad p^{-1}([x]) = \{x, -x\} \subset S^2.$$

Then the **real projective plane** $\mathbb{RP}^2$ has a quotient topology induced by the projection $p : S^2 \longrightarrow \mathbb{RP}^2$.

Theorem 4.2.5. *For a quotient map $p : X \longrightarrow X/\sim$, the followings are equivalent :*

(1) *p is open.*

(2) *If $A \subset X$ is open, then $\{x \in X \,|\, x \sim y, y \in A\}$ is open in X.*

(3) *If $A \subset X$ is closed, then $X - \{x \in X \,|\, x \sim y, y \in X - A\}$ is closed in X.*

Proof. For any subset $A \subset X$, $p^{-1}(p(A)) = \{x \in X \,|\, x \sim y, y \in A\}$.

(1) $\Rightarrow$ (2) If A is open, then $p(A)$ is open in $X/\sim$.
Since p is continuous

$$p^{-1}\big(p(A)\big) = \big\{x \in X \,|\, x \sim y, y \in A\big\}$$

is open in X.

(2) $\Rightarrow$ (1) If A is open, then $p^{-1}(p(A))$ is open in X. Since $X/\sim$ has the identification topology by p, $p(A)$ is open in $X/\sim$.

(2) $\Rightarrow$ (3) Since $X - \{x \in X \,|\, x \sim y,\ y \in X - A\} = X - p^{-1}\big(p(X - A)\big)$ if A is closed, then $X - A$ is open.

By (2) $p^{-1}\big(p(X-A)\big)$ is open in X.
Thus $X - p^{-1}(p(X-A))$ is closed in X.

(3) $\Rightarrow$ (2) If A is open, then $X - A$ is closed.
By (3) $X - p^{-1}\big(p(A)\big)$ is closed. Thus $p^{-1}\big(p(A)\big)$ is open. $\square$

Theorem 4.2.6. *Consider the following diagram:*

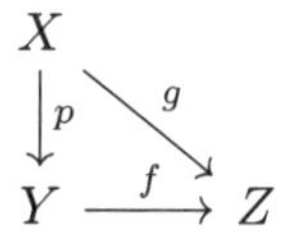

Let $f \circ p = g$ and p be a quotient map. Then f is continuous iff g is continuous.

Proof. ($\Rightarrow$) If f is continuous, then $f \circ p = g$ is continuous.
($\Leftarrow$) If $g = f \circ p$ is continuous and W is open in Z, then

$$g^{-1}(W) = (f \circ p)^{-1}(W) = p^{-1}(f^{-1}(W))$$

is open in X. Since Y has a quotient topology by p, $f^{-1}(W)$ is open in Y. Thus f is continuous. $\square$

4.2. Exercises

1. Let $f : X \longrightarrow Y$ be a continuous, surjective map, and

$$\widetilde{X} = \{f^{-1}(y) \,|\, y \in Y\}$$

have the quotient topology induced by X and f. Prove :

(1) If Y is Hausdorff, then so is $\widetilde{X}$.

(2) The induced map $\widetilde{f} : \widetilde{X} \longrightarrow Z$ given by $\widetilde{f}(x) = f(x)$ is a homeomorphism iff f is a quotient map.

2. Let $\{X_\alpha \mid \alpha \in I\}$ be a family of topological spaces and $p_\alpha : X_\alpha \longrightarrow Y$ be a map for each $\alpha \in I$, where Y is a set. Prove:

(1) There exists a unique finest topology $\mathcal{T}$ on Y such that each map p_α is continuous.

(2) A map $f : (Y, \mathcal{T}) \longrightarrow Z$ is continuous iff the map $f \circ p_\alpha$ is continuous for each $\alpha \in I$.

3. For each $\alpha \in I$, let $f_\alpha : X_\alpha \longrightarrow Y_\alpha$ be a continuous, open, surjective map. Show the product

$$\prod_{\alpha\in I} f_\alpha : \prod_{\alpha\in I} X_\alpha \longrightarrow \prod_{\alpha\in I} Y_\alpha$$

is a quotient map.

Chapter 5

Connected Spaces and Compact Spaces

5.1 Connected Spaces

Definition 5.1.1. A topological space X is **connected** if X and ϕ are only open and closed sets in X.

Theorem 5.1.2. *X is connected iff X is not a union of disjoint nonempty open (closed) subsets.*

Proof. Suppose that $X = U_1 \cup U_2$ is a disjoint union of nonempty open subsets U_1 and U_2. Then U_1 and U_2 are closed.

If $U_1 \neq X$ since if $U_1 = X$, then $U_2 = \emptyset$.

If $U_1 \neq \emptyset$ since if $U_1 = \emptyset$, then $U_2 = X$. This contradicts to the connectedness of X.

Conversely, suppose U is an open and closed subset of X such that $U \neq \emptyset$ and $U \neq X$. Then $X - U$ is a closed and open subset of X, and $X - U \neq X$ and $X - U \neq \emptyset$.
Thus $X = U \cup (X - U)$ is a union of nonempty disjoint open subsets. Contradict! □

Definition 5.1.3. Let $(X, \mathcal{T})$ be a topological space and $A \subset X$. A is **connected** if $(A, \mathcal{T}_A)$ is connected.

Theorem 5.1.4. *Let A_0 and $\{A_i \mid i \in I\}$ be a family of connected subspaces of a space X such that $A_0 \cap A_i \neq \emptyset$, $\forall i \in I$. Then*

$$A = A_0 \bigcup \Big(\bigcup_{i \in I} A_i\Big)$$

is connected.

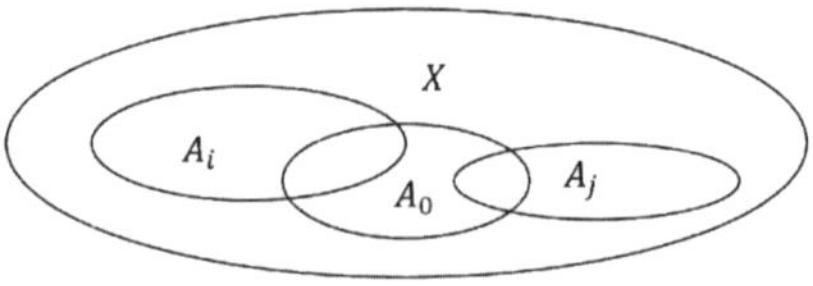

Proof. Suppose $A = U_1 \cup U_2$ is a union of disjoint open sets U_1 and U_2. For each $i \in I$, $A_i = (U_1 \cap A_i) \cup (U_2 \cap A_i)$ is a union of disjoint open sets.

Since A_i is connected, either $U_1 \cap A_i = \emptyset$ or $U_2 \cap A_i = \emptyset$.

Similarly, either $U_1 \cap A_0 = \emptyset$ or $U_2 \cap A_0 = \emptyset$.

If $U_2 \cap A_0 = \emptyset$, then $U_1 \cap A_0 = A_0$ and $A_0 \subset U_1$. Since for each $i \in I$ $A_0 \cap A_i \neq \emptyset$ and $U_1 \cap A_i \neq \emptyset$,

$$U_2 \cap A_i = \emptyset \quad \forall i \in I.$$

Then $A_i = U_1 \cap A_i$ $(i \in I)$ and $A_i \subset U_1$ $(i \in I)$. So,

$$U_1 = A_0 \bigcup \Big(\bigcup_{i \in I} A_i\Big) \quad \text{and} \quad U_2 = \emptyset.$$

Thus A is connected. □

Example 5.1.5. (1) $\mathbb{R}^n$ is connected.

(2) $\mathbb{R}^n - \{0\}$ is connected if $n > 1$.

(3) $\mathbb{R}^1 - \{0\}$ is not connected.

(4) In $\mathbb{R}^2$,

$$A_n = \mathbb{R} \times \{n\} \text{ is connected, } n \in \mathbb{Z},$$
$$A_* = \{0\} \times \mathbb{R} \text{ is connected.}$$

Then $\bigcup_{n\in\mathbb{Z}} A_n$ is not connected, but $A = A_* \cup \left(\cup_{n\in\mathbb{Z}} A_n\right)$ is connected, because $A_* \cap A_n = \{(0, n)\} \neq \emptyset$ for each n.

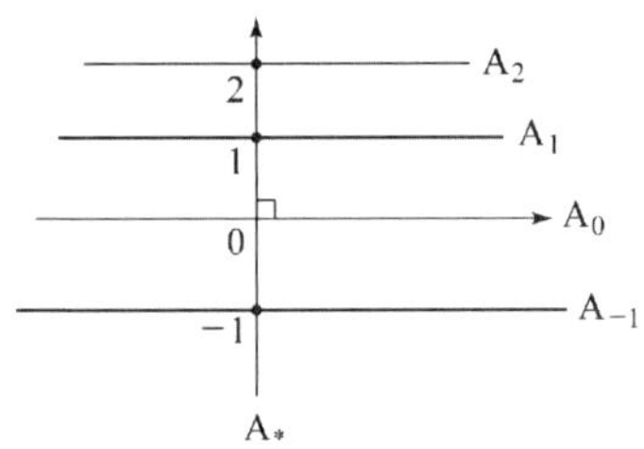

Theorem 5.1.6. *If $f : X \longrightarrow Y$ is continuous and surjective, and X is connected, then Y is connected.*

Proof. If $Y = V_1 \cup V_2$ is a union of disjoint open subsets, then

$$X = f^{-1}(V_1) \cup f^{-1}(V_2)$$

is a union of disjoint open subsets. Since X is connected, either $f^{-1}(V_1) = \emptyset$ or $f^{-1}(V_2) = \emptyset$. Since f is onto $f(f^{-1}(V_i)) = V_i$, $i = 1, 2$. Thus either $V_1 = \emptyset$ or $V_2 = \emptyset$. □

Corollary 5.1.7. *Let $f : X \longrightarrow Y$ be continuous and X connected. Then $f(X)$ is connected.*

Example 5.1.8. For $n > 1$, let

$$f : \mathbb{R}^n - \{0\} \longrightarrow S^{n-1}, \quad f(x) = \frac{x}{|x|}.$$

Then f is continuous and onto, $\mathbb{R}^n - \{0\}$ and S^{n-1} are connected.

Definition 5.1.9. A topological space X is called **path connected** if for any two points $x_0, x_1 \in X$, there is a continuous map $\alpha : I \longrightarrow X$ such that $\alpha(0) = x_0$ and $\alpha(1) = x_1$.

Theorem 5.1.10. *If X is path connected, then X is connected.*

Proof. Let $X = U_0 \cup U_1$ be a disjoint union of nonempty open subsets U_0 and U_1 of X. Choose points $x_0 \in U_0$ and $x_1 \in U_1$.

Since X is path connected, there is a continuous map $\alpha : I \longrightarrow X$ such that $\alpha(0) = x_0$ and $\alpha(1) = x_1$. While

$$\alpha^{-1}(U_0) \cup \alpha^{-1}(U_1) = \alpha^{-1}(U_0 \cup U_1) = \alpha^{-1}(X) = I$$

is a disjoint union of nonempty open subsets.

We have a contradiction since $I = [0, 1]$ is connected. □

Example 5.1.11. The topologist sine curve

$$X = \{(0, y) \in \mathbb{R}^2 \,|\, -1 \leq y \leq 1\} \bigcup \{(x, \sin\frac{1}{x}) \in \mathbb{R}^2 \,|\, 0 < x \leq 1\}$$

is connected but not path connected.

5.1. Exercises

1. Let $\{X_\alpha \,|\, \alpha \in I\}$ be a family of nonempty topological spaces. Then prove that the product $\prod_{\alpha \in I} X_\alpha$ is connected iff X_α is connected for each $\alpha \in I$.

2. Let $\mathfrak{F}$ be a family of path connected subsets of X with a nonempty intersection. Then prove that the union

$$Z = \cup\{F \,|\, F \in \mathfrak{F}\}$$

is path connected.

3. Prove Example 5.1.11.

5.2 Compact Spaces

Definition 5.2.1. (1) A family $\{U_\alpha \,|\, \alpha \in I\}$ of open subsets of X such that $X = \cup_{\alpha\in I} U_\alpha$ is called an **open cover of** X.

(2) A space X is **compact** if every open cover of X has a finite subcover, that is, if $\{U_\alpha \,|\, \alpha \in I\}$ is an open cover of X, then there is a finite subset $I_1 \subset I$ such that $\{U_\alpha \,|\, \alpha \in I_1\}$ is a cover of X,

$$\bigcup_{\alpha\in I} U_\alpha = X = \bigcup_{\alpha\in I_1} U_\alpha.$$

Example 5.2.2. (1) **(Heine-Borel Theorem)** *A subset $X \subset \mathbb{R}^n$ is compact iff X is bounded and closed.*

(2) Every open subset of $\mathbb{R}^n$ is not compact.

(3) $X = \{0\} \cup \{\frac{1}{n} \,|\, n \in \mathbb{N}\}$ is a compact subset of $\mathbb{R}^1$, but $\{\frac{1}{n} \,|\, n \in \mathbb{N}\}$ is not compact.

Definition 5.2.3. A space X has the **finite intersection property** (f.i.p) iff for every family $\{F_\alpha \,|\, \alpha \in I\}$ of closed subsets of X if

$$\bigcap_{\alpha\in I_1} F_\alpha \neq \emptyset$$

for every finite subset $I_1 \subset I$, then $\cap_{\alpha\in I} F_\alpha \neq \emptyset$.

Theorem 5.2.4. *A space X is compact iff X has the f.i.p.*

Proof. There is a one-to-one correspondence between $\mathfrak{U} = \{U \,|\, U:\ \text{open}\}$ and $\mathfrak{F} = \{F \,|\, F:\ \text{closed}\}$ by $U = X - F$ and $X - U = F$. Moreover,

$$\begin{aligned}\bigcup_{\alpha\in I} U_\alpha = X \quad \text{iff} \quad & \bigcap_{\alpha\in I}(X - U_\alpha) = \emptyset \\ & = \Big(\bigcap_{\alpha\in I} U_\alpha^c\Big) \\ & = \Big(\bigcup_{\alpha\in I} U_\alpha\Big)^c.\end{aligned}$$

Suppose $\mathfrak{F} = \{F_\alpha \,|\, \alpha \in I\}$ is a family of closed sets such that for every finite $I_1 \subset I$, $\cap_{\alpha \in I_1} F_\alpha \neq \emptyset$ but $\cap_{\alpha \in I} F_\alpha = \emptyset$.
Then $\cup_{\alpha \in I} F_\alpha^c = X$ and $\{F_\alpha^c \,|\, \alpha \in I\}$ is an open cover of X, but $\cup_{\alpha \in I_1} F_\alpha^c \neq X$ for every finite $I_1 \subset I$.

Conversely, if $\mathfrak{U} = \{U_\alpha \,|\, \alpha \in I\}$ is an open cover of X such that $\cup_{\alpha \in I_1} U_\alpha \neq X$ for any finite $I_1 \subset I$. Then

$$\cup_{\alpha \in I} U_\alpha = X, \quad \cap_{\alpha \in I} U_\alpha^c = \emptyset$$

and $\cap_{\alpha \in I_1} U_\alpha^c \neq \emptyset$ for any finite $I_1 \subset I$. □

Theorem 5.2.5. *If X is compact and $A \subset X$ is closed, then A is compact.*

Proof. If $\mathfrak{F} = \{F_\alpha \,|\, \alpha \in I\}$ is a family of closed subsets of A with $\cap_{\alpha \in I_1} F_\alpha \neq \emptyset$ for any finite $I_1 \subset I$. Since X is compact and $\mathfrak{F}$ is a family of closed sets in X with $\cap_{\alpha \in I_1} F_\alpha \neq \emptyset$ for any finite $I_1 \subset I$.
Thus $\cap_{\alpha \in I} F_\alpha \neq \emptyset$, and A is compact. □

Theorem 5.2.6. *Let X be compact and $A \subset X$ be an infinite subset. Then A has a limit point.*

Proof. Suppose that A has no limit point. For each $a \in A$ there is $U(a)$ is a nbd of a such that $U(a) \cap A = \{a\}$. Then $\{a\}$ is an open subset of A. Since $A' = \emptyset$, $\bar{A} = A \cup A' = A$ and A is closed.

By Theorem 5.2.5, A is compact.

However, A is infinite and $\{a\}$ is open in A for all $a \in A$. Thus we have a contradiction and so A has a limit point. □

Theorem 5.2.7. *Let $f : X \longrightarrow Y$ be continuous and surjective. If X is compact, then Y is compact.*

Proof. If $\{V_\alpha \,|\, \alpha \in I\}$ is an open cover of Y, then $\{f^{-1}(V_\alpha) \,|\, \alpha \in I\}$ is an open cover of X. Then there is $I_1 \subset I$ is finite such that $\{f^{-1}(V_\alpha) \,|\, \alpha \in I_1\}$ is a cover of X. Since f is onto, $f(f^{-1}(V_\alpha)) = V_\alpha$ for each $\alpha \in I$. Then

$$Y = f(x) = f\Big(\bigcup_{\alpha \in I_1} f^{-1}(V_\alpha)\Big) = \bigcup_{\alpha \in I_1} f\Big(f^{-1}(V_\alpha)\Big) = \bigcup_{\alpha \in I_1} V_\alpha.$$

Thus Y is compact. □

Corollary 5.2.8. (1) *If $f : X \longrightarrow Y$ is continuous and X is compact, then $f(X)$ is compact.*

(2) *If $f : X \longrightarrow \mathbb{R}$ is continuous on a compact space X, there are x_1, $x_2 \in X$ such that $f(x_1) \le f(x) \le f(x_2)$ for all $x \in X$.*

Proof. (2) Since $f(X)$ is closed and bounded, there are $\text{lub} f(X)$ and $\text{glb} f(X)$ in $f(X) \subset \mathbb{R}$. There are x_1 and $x_2 \in X$ such that

$$f(x_1) = \text{glb} f(x), \quad f(x_2) = \text{lub} f(x).$$

Thus $f(x_1) \le f(x) \le f(x_2)$ for all $x \in X$. □

Theorem 5.2.9 (Tychonoff). *If $\{X_\alpha \,|\, \alpha \in I\}$ is a family of compact spaces, then $X = \prod_{\alpha \in I} X_\alpha$ is compact.*

Proof. Suppose that $\mathfrak{F} = \{F_\beta \,|\, \beta \in J\}$ is a family of closed subsets in X such that $\cap_{\beta \in J_1} F_\beta \ne \emptyset$ for any finite $J_1 \subset J$. Let $\widetilde{\mathfrak{F}}$ be a maximal family of closed subsets in X such that $\mathfrak{F} \subset \widetilde{\mathfrak{F}}$ and any finite intersection is nonempty. For each $\alpha \in I$ let

$$\widetilde{\mathfrak{F}}_\alpha = \{\overline{p_\alpha(F)} \,|\, F \in \widetilde{\mathfrak{F}}\}.$$

Then $\widetilde{F_\alpha}$ is a family of closed subsets of X_α and any finite intersection is nonempty. Since X_α is compact there is $X_\alpha \cap \overline{p_\alpha(F)} \subset X_\alpha$. Then $x = (x_\alpha)_{\alpha \in I} \in X$. For each $F_\beta \in \widetilde{\mathfrak{F}}$, $x \in \overline{F_\beta} = F_\beta$. Thus

$$x \in \cap_{F \in \widetilde{\mathfrak{F}}} F \subset \cap_{F \in \mathfrak{F}} F \quad \text{and} \cap_{F \in \mathfrak{F}} F \neq \emptyset.$$

Hence X is compact. □

Definition 5.2.10. A map $f : (X, d_X) \longrightarrow (Y, d_Y)$ between metric spaces is **uniformly continuous** iff for any $\varepsilon > 0$ there is $\delta > 0$ such that $d_Y(f(x_1), f(x_2)) < \varepsilon$ if $d_X(x_1, x_2) < \delta$.

Lemma 5.2.11. *Let (X, d) be a compact metric space. If $\{U_\alpha \mid \alpha \in I\}$ is an open cover of X, then there is $\delta > 0$ such that for any point $x \in X$ and a δ-ball $B(x, \delta) \subset U_\alpha$ for some $\alpha \in I$.* (The δ is called **Lebesgue number** of the cover $\{U_\alpha \mid \alpha \in I\}$).

Proof. Since X is compact, there is a finite subcover of X,

$$\{U_1, \cdots, U_n\} \subset \{U_\alpha\}_{\alpha \in I}.$$

For each $i \in \{1, 2, \cdots, n\}$, $X - U_i = F_i$ is closed and compact. If $X \in \{U_\alpha\}_{\alpha \in I}$, then Theorem 5.2.7 holds.

If $X \notin \{U_\alpha\}_{\alpha \in I}$, then we define a map $f : X \longrightarrow \mathbb{R}$ by

$$f(x) = \frac{1}{n} \sum_{i=1}^{n} d(x, F_i), \quad x \in X$$

where $d(x, F_i) := \min\{d(x, y) \mid y \in F_i\}$.
Then $d(x, \cdot) : F_i \longrightarrow \mathbb{R}$ is continuous for each $i = \{1, \cdots, n\}$.
Thus f is continuous and $f(x) > 0$ for all $x \in X$.

Indeed, if $x \in X$, then $x \in U_i$ for some $i \in \{1, \cdots, n\}$. So, $x \notin F_i$ and $d(x, F_i) > 0$. In fact, there is $\varepsilon > 0$ such that $B(x, \varepsilon) \subset U_i$ and

$F_i = X - U_i \subset X - B(x, \varepsilon)$. Then $d(x, F_i) \geq \varepsilon$ and $f(x) > \frac{\varepsilon}{n} > 0$. Thus $f(x)$ is compact.

Let $\delta := \min\{f(x) \,|\, x \in X\}$. Then $\delta > 0$, and $x_0 \in X$, $B(x_0, \delta) \subset U_k$ for some $k \in \{1, \cdots, n\}$.

Indeed, let $d(x_0, F_k) = \max\{d(x_0, F_i) \,|\, i = 1, 2, \cdots, n\}$. For any $x \in B(x_0, \delta)$,

$$d(x_0, x) < \delta \leq f(x_0) = \frac{1}{n}\sum_{i=1}^{n} d(x_0, F_i) \leq d(x_0, F_k).$$

Then $x \notin F_k = X - U_k$ and $x \in U_k$.
Thus $B(x_0, \delta) \subset U_k$. □

Theorem 5.2.12 (Uniform Continuity Theorem). *If (X, d_X) is compact, then a continuous map $f : (X, d_X) \longrightarrow (Y, d_Y)$ is uniformly continuous.*

Proof. For any $\varepsilon > 0$, choose an open covering of Y, $\{B(y, \frac{\varepsilon}{2}) \,|\, y \in Y\}$. Then $\{f^{-1}(B(y, \frac{\varepsilon}{2})) \,|\, y \in Y\}$ is an open covering of X. Since (X, d_X) is a compact metric space, by Lemma 5.2.11, there is $\delta > 0$ such that if $d_X(x_1, x_2) < \delta$, then x_1, $x_2 \in f^{-1}(B(y, \frac{\varepsilon}{2}))$ for some $y \in Y$. Then $f(x_1)$, $f(x_2) \in B(y, \frac{\varepsilon}{2})$ and

$$d_Y(f(x_1), f(x_2)) \leq d_Y(f(x_1), y) + d_Y(y, f(x_2)) < \varepsilon.$$

Thus f is uniformly continuous. □

Example 5.2.13. If f is uniformly continuous, then f is continuous.

(1) The map $f_1 : [0, 1) \longrightarrow \mathbb{R}$ defined by $f_1(x) = 1 - x$, $x \in [0, 1)$ is uniformly continuous.

(2) The map $f_2 : [0, 1) \longrightarrow \mathbb{R}$ defined by $f_2(x) = \frac{1}{1-x}$, $x \in [0, 1)$ is continuous but not uniformly continuous.

(3) If $f_3 : [0,1] \longrightarrow \mathbb{R}$ is continuous, then f_3 is uniformly continuous, by Theorem 5.2.7, since $[0,1]$ is compact.

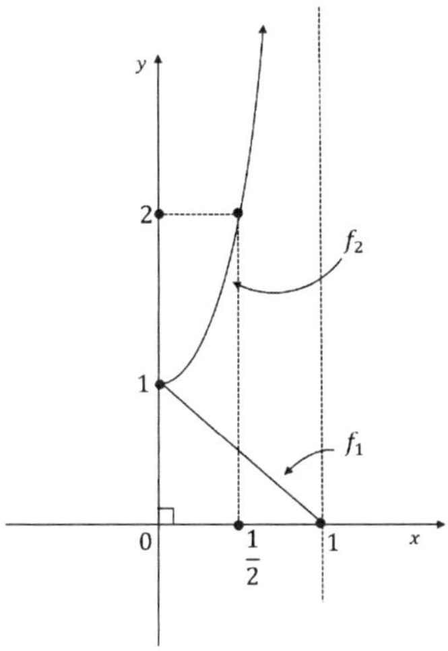

5.2. Exercises

1. Prove that every compact subset of a Hausdorff space is closed.

2. Show that the unit interval $[0,1]$ in $(\mathbb{R}^1, \mathcal{T}_d)$ is compact.

3. Prove that compactness and connectedness are topological invariants.

4. Let $f : X \longrightarrow Y$ be a map, and Y compact Hausdorff. Then show that f is continuous iff the graph of f,

$$G(f) = \{(x, f(x)) \,|\, x \in X\},$$

is closed in $X \times Y$.

5. Let A_1 and A_2 be disjoint compact subsets of a Hausdorff space X. Then show that there are disjoint open sets U_1 and U_2 such that $A_1 \subset U_1$ and $A_2 \subset U_2$.

6. Prove that a compact subset of a metric space is closed and bounded.

7. Let X be a set, and $\mathcal{T} = \{X - F \mid F \text{ is a finite subset of } X\} \cup \{\emptyset\}$. Then show that

(1) $\mathcal{T}$ is a topology on X.

(2) $(X, \mathcal{T})$ is a compact space.

Chapter 6

Classification of Topological Spaces

6.1 Separation Axioms

Definition 6.1.1. Let X be a topological space.

(1) X is a $\mathbf{T_0}$**-space** if for $x_1 \neq x_2 \in X$, there is a open subset $U \subset X$ such that U contains one of x_1 and x_2.

(2) X is a $\mathbf{T_1}$**-space** if for $x_1 \neq x_2 \in X$, there are two open subsets U_1, U_2 such that $x_1 \in U_1$ and $x_2 \notin U_1$, and $x_1 \notin U_2$ and $x_2 \in U_2$.

(3) X is a $\mathbf{T_2}$**-space** if for $x_1 \neq x_2 \in X$, there are open subsets U_1, U_2 in X such that $x_1 \in U_1$, $x_2 \in U_2$, and $U_1 \cap U_2 = \emptyset$. A T_2-space is also called a **Hausdorff space**.

Example 6.1.2. (1) T_2-space $\Longrightarrow$ T_1-space $\Longrightarrow$ T_0-space.

(2) If X contains more than one elements, and $\mathcal{T} = \{X, \emptyset\}$, then $(X, \mathcal{T})$ is not a T_0-space.

(3) If $X = \{x_1, x_2\}$ and $\mathcal{T} = \{X, \emptyset, \{x_1\}\}$, then $(X, \mathcal{T})$ is T_0-space but T_1-space.

(4) If X contains infinite elements and $\mathcal{T} = \{\emptyset\} \cup \{X - F \mid F \text{ is finite}\}$, then $(X, \mathcal{T})$ is T_1 but not T_2.

(5) If (X, d) is a metric space, then X is T_2.

Theorem 6.1.3. *A space X is T_1 iff each point subset of X is closed, i.e., $\{x\}$ is closed for each point $x \in X$.*

Proof. ($\Longrightarrow$) For each point $x_1 \in X$, and $x \in X - \{x_1\}$, there is a nbd $U(x)$ of x such that $x \in U(x)$ and $x_1 \notin U(x)$. Then the union of such a nbd $U(x)$ of x

$$X - \{x_1\} = \bigcup_{\substack{x \in X - \{x_1\} \\ x_1 \notin U(x)}} U(x)$$

is open in X.
Thus $\{x_1\}$ is closed.

($\Longleftarrow$) For $x_1 \neq x_2 \in X$, let $U_1 = X - \{x_2\}$, $U_2 = X - \{x_1\}$. Then U_1 and U_2 are open, and $x_1 \in U_1$, $x_2 \notin U_1$ and $x_1 \notin U_2$, $x_2 \in U_2$.
Thus X is T_1. □

Theorem 6.1.4. *Let X be T_2 and $K \subset X$ compact. Then K is closed.*

Proof. Show $X - K$ is open. For each point $x \in X - K$ and $y \in K$, there exist $U(x)$ and $V(y)$ which are the neighborhoods of x and y, respectively, with $U(x) \cap V(y) = \emptyset$.

If $\{V(y) \cap K \mid y \in K\}$ is an open cover of K, then there is a finite subcover $\{V(y_1) \cap K, \cdots, V(y_n) \cap K\}$. So,

$$K = \bigcup_{i=1}^{n} (V(y_i) \bigcap K) \subset \bigcup_{i=1}^{n} V(y_i) := V(K).$$

Let $U(x) = \cap_{i=1}^{n} U(y_i)$. Then $U(x)$ is a nbd of x. Then

$$\begin{aligned} U(x) \cap K &= \Big(\bigcap_{j=1}^{n} U(y_j)\Big) \bigcap \Big(\bigcup_{i=1}^{n} V(y_i)\Big) \\ &= \bigcup_{i=1}^{n} \Big[\Big(\bigcap_{j=1}^{n} U(y_j)\Big) \bigcap V(y_i)\Big] \\ &\subset \bigcup_{i=1}^{n} \Big(U(y_i) \bigcap V(y_i)\Big) = \emptyset. \end{aligned}$$

Thus $U(x) \subset X - K$ and $U(x) \cap V(K) = \emptyset$. Therefore $X - K$ is open. □

Theorem 6.1.5. *Let X be a T_2-space and K_1, K_2 are disjoint compact subsets of X. Then there exist U a nbd of K_1 and V a nbd of K_2 such that $U \cap V = \emptyset$.*

Proof. As in the proof of Theorem 6.1.4, we can find for each point $x \in K_1$ and K_2, $U(x)$ a nbd of x and $V(x)$ a nbd of K_2 such that $U(x) \cap V(x) = \emptyset$. The set $\{U(x) \cap K_1 \mid x \in K_1\}$ is an open cover of K_1. Since K_1 is compact, there is a finite subcover $\{K_1 \cap U(x_i) \mid i = 1, 2, \cdots, n\}$ of $\{K_1 \cap U(x) \mid x \in K_1\}$.

Let $U = \cup_{i=1}^{n} U(x_i)$ and $V = \cap_{i=1}^{n} V(x_i)$. Then U and V are open, and $K_1 = \cup_{i=1}^{n} (K_1 \cap U(x_i)) \subset \cup_{i=1}^{n} U(x_i) = U$ since $K_2 \subset V(x_i)$, $i = 1, \cdots, n$, $K_2 \subset \cap_{i=1}^{n} V(x_i) = V$. Moreover,

$$\begin{aligned} U \cap V &= \Big(\bigcup_{i=1}^{n} U(x_i)\Big) \bigcap \Big(\bigcap_{i=1}^{n} V(x_i)\Big) \\ &= \bigcup_{i=1}^{n} \Big[U(x_i) \bigcap \Big(\bigcap_{j=1}^{n} V(x_j)\Big)\Big] \\ &\subset \bigcup_{i=1}^{n} \Big(U(x_i) \bigcap V(x_j)\Big) = \emptyset. \end{aligned}$$

Thus we complete the proof. □

Theorem 6.1.6. *Let $f : X \longrightarrow Y$ be continuous and bijective. If X is compact and Y is T_2, then f is a homeomorphism.*

Proof. Show $f^{-1} : Y \longrightarrow X$ is continuous. If F is closed X, then F is compact and $f(F)$ is $f(F)$ is compact in Y. Since Y is T_2, $f(F)$ is closed in Y, and $(f^{-1})^{-1}(F) = f(F)$ is closed in Y by Theorem 6.1.4.
Thus f^{-1} is continuous. □

Example 6.1.7. (1) Let $(X = \mathbb{R}^1, \mathfrak{D})$, $\mathfrak{D}$ a discrete topology and $(Y = \mathbb{R}^1, d)$ the usual space. Then $f : X \longrightarrow Y$ given by $f(x) = x$ for all $x \in X$ is continuous, bijective, but not a homeomorphism, because X is not compact.

(2) Let $X = \{x_1, x_2\}$ and $Y = \{y_1, y_2\}$. Let $\mathcal{T}_X = \mathfrak{D}$ be a discrete topology, $\mathcal{T}_Y = \{Y, \phi, \{y_2\}\}$. The map $f : X \longrightarrow Y$, $f(x_i) = y_i$, $i = 1, 2$ is continuous and bijective but not a homeomorphism, since Y is not T_2.

Definition 6.1.8. A sequence $< x_n >$ in a space X **converges to** a point $x \in X$ if for any nbd U of x there is n_0 such that $x_n \in U$ for all $n > n_0$.

Theorem 6.1.9. *If X is T_2 and a sequence $< x_n >$ converges to x and y, then $x = y$.*

Proof. If $x \neq y$, then there are U a nbd of x and V a nbd of y such that $U \cap V = \emptyset$. Since $< x_n >$ converges to x, there is n_1 such that $x_n \in U$ for $n > n_1$, and since $< x_n >$ converges to y, there is n_2 such

that $x_n \in V$ for $n > n_1$.
Thus $x_n \in U \cap V$ for any $n > \max\{n_1, n_2\}$.
This contradicts to $U \cap V = \emptyset$. □

Example 6.1.10. If $(X, \mathcal{T})$ is an indiscrete space, then every sequence converges to every point in X.

Definition 6.1.11. (1) A topological space X is **regular** if for each close F, $x \in X - F$ there are open U_1 and U_2 such that $F \subset U_1$, $x \in U_2$ and $U_1 \cap U_2 = \emptyset$.

(2) X is **$\mathbf{T_3}$-space** if X is T_1 and regular.

(3) X is **normal** if for each disjoint closed subsets F_1 and F_2 there are disjoint open subsets U_1 and U_2 such that $F_i \subset U_i$, $i = 1, 2$.

(4) X is **$\mathbf{T_4}$-space** if X is T_1 and normal.

Example 6.1.12. (1) T_4-space $\Rightarrow$ T_3-space $\Rightarrow$ T_2-space $\Rightarrow$ T_1-space $\Rightarrow$ T_0-space.

(2) Construct a space which is T_2 but not T_3. Let

$$\begin{aligned}\mathbb{R}^2_{\geq 0} &= \{(x_1, x_2) \in \mathbb{R}^2 \mid x_2 \geq 0\},\\ \mathbb{R}^2_{>0} &= \{(x_1, x_2) \in \mathbb{R}^2 \mid x_2 > 0\},\\ \mathbb{R}^2_0 &= \{(x_1, 0) \in \mathbb{R}^2\}.\end{aligned}$$

Then, $\mathbb{R}^2_{\geq 0} = \mathbb{R}^2_{>0} \cup \mathbb{R}^2_0$.

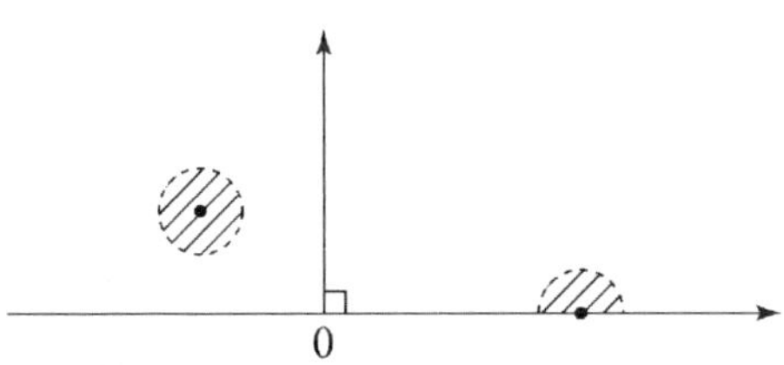

Construct a topology on $\mathbb{R}^2_{\geq 0}$ as follows:

Let $B(x,r) = \{y \in \mathbb{R}^2 \,|\, |x-y| < r\}$, $x \in \mathbb{R}^2$, $r > 0$,

$$\mathcal{B}_1 = \Big\{B(x,r) \,|\, x = (x_1, x_2) \in \mathbb{R}^2_{>0}, 0 < r < x_2\Big\},$$

$$\mathcal{B}_2 = \Big\{\big[B(x,r) \bigcap \mathbb{R}^2_{>0}\big] \bigcup \, \{x\} \,|\, x \in \mathbb{R}^2_0\Big\},$$

and $\mathcal{B} := \mathcal{B}_1 \cup \mathcal{B}_2$.

Then $\mathcal{B}$ is a basis for a topology $\mathcal{T}_\mathcal{B}$ of $\mathbb{R}^2_{\geq 0}$, and $(\mathbb{R}^2_{\geq 0}, \mathcal{T}_\mathcal{B})$ is T_2 but not T_3.

Indeed, clearly, $(\mathbb{R}^2_{\geq 0}, \mathcal{T}_\mathcal{B})$ is T_2.

Consider $0 \in \mathbb{R}^2_{\geq}$ and $\mathbb{R}^2_0 - \{0\}$ closed in $\mathbb{R}^2_{\geq 0}$. Let U be a nbd of 0. Then $(B(0,r) \cap \mathbb{R}^2_{\geq 0}) \cup \{0\} \subset U$, and (any nbd of $\frac{r}{2} \in \mathbb{R}^2_0 - \{0\}) \cap U \neq \emptyset$. Thus 0, $\mathbb{R}^2_0 - \{0\}$ can not separate in $(\mathbb{R}^2_{\geq}, \mathcal{T}_B)$, i.e., $\mathbb{R}^2_{\geq 0}$ is neither regular nor T_3.

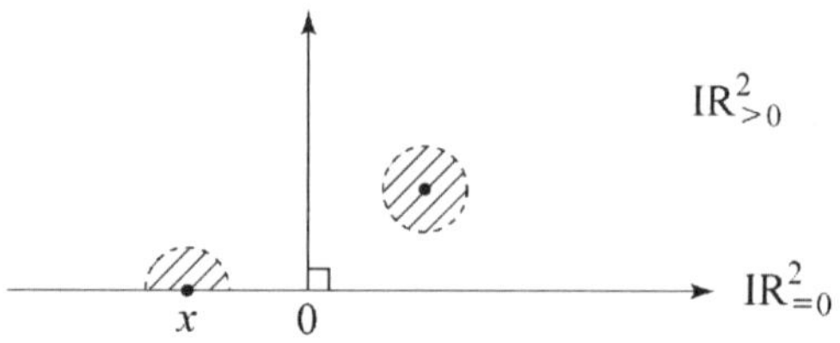

(3) If X is a compact T_2-space, then X is T_4. If F_1, F_2 are disjoint closed subsets in X, then F_1, F_2 are disjoint compact subsets in X. By Theorem 6.1.3, F_1 and F_2 have disjoint open nbds in X.

Theorem 6.1.13. *If X is regular and $F \subset X$ closed, $x \in X - F$, then there are open U_1 and U_2 such that $F \subset U_1$, $x \in U_2$, $\overline{U_1} \cap \overline{U_2} = \emptyset$.*

Proof. Since X is regular, there are disjoint open V_1 and V_2 such that $F \subset V_1$ and $x \in V_2$. Since $X - V_2$ is closed and $x \notin X - V_2$, there are disjoint open W_1 and W_2 such that $X - V_2 \subset W_1$ and $x \in W_2$.

Let $U_1 = V_1$, $U_2 = W_2$. Then $F \subset U_1$ and $x \in U_2$ and

$$\overline{U_1} = \overline{V_1} \subset \overline{X - V_2} = X - V_2,$$

$$\overline{U_2} = \overline{W_2} \subset \overline{X - W_1} = X - W_1 \subset V_2$$

and so $\overline{U}_1 \cap \overline{U_2} \subset (X - V_2) \cap V_2 = \emptyset$. □

Theorem 6.1.14. *If X is normal and F_1, F_2 are disjoint closed, then there are disjoint open U_1, U_2 such that $F_i \subset U_i$, $i = 1, 2$, $\overline{U_1} \cap \overline{U_2} = \emptyset$.*

Proof. The proof is similar to Theorem 6.1.13. □

Theorem 6.1.15. *Every metric space (X, d) is normal and T_4.*

Proof. (1) For each point $x \in X$, let $y \in X - \{x\}$ and $d(x, y) = r > 0$. Then $B(y, r) \subset X - \{x\}$. Thus $X - \{x\}$ is open and $\{x\}$ is closed. Therefore, X is T_1.

(2) Let F_1 and F_2 be disjoint closed in X. Define two functions by

$$d_1 : X \longrightarrow \mathbb{R}, \quad d_1(x) := \mathrm{glb}\{d(x, y_1) \,|\, y_1 \in F_1\},$$
$$d_2 : X \longrightarrow \mathbb{R}, \quad d_2(x) := \mathrm{glb}\{d(x, y_2) \,|\, y_2 \in F_2\},$$

where glb stands for the greatest lower bound.

Let $U_1 = \{x \in X \,|\, d_1(x) < d_2(x)\} = (d_1 - d_2)^{-1}(-\infty, 0)$,
$U_2 = \{x \in X \,|\, d_1(x) > d_2(x)\} = (d_1 - d_2)^{-1}(0, \infty)$.

Then $U_1 \cap U_2 = \emptyset$.

Claim. (i) *$F_1 \subset U_1$ and $F_2 \subset U_2$,*

(ii) *U_1, U_2 are open.*

(i) If $x \in F_1$, then $d_1(x) = 0$ and $d_2(x) > 0$.

If $d_2(x) = 0 = \text{glb}\{d(x, y) \,|\, y \in F_2\}$, then for any $\varepsilon > 0$, there is $y_2 \in F_2$ such that $y_2 \in B(x, \varepsilon)$. Then x is a limit point of F_2. So $x \in \overline{F_2} = F_2$.

Contradict! since $F_1 \cap F_2 = \emptyset$.

Thus $F_1 \subset U_1$.

Similarly, we can prove $F_2 \subset U_2$.

(ii) Show U_1 is open.

For any point $x_1 \in U_1$, let $a := d_2(x_1) - d_1(x_1) > 0$.

Then $B(x_1, \frac{a}{3}) \subset U_1$.

Indeed: For any point $x \in B(x_1, \frac{a}{3})$,

$$\begin{aligned} d_1(x) - d_2(x) &= (d_1(x) - d_1(x_1)) + (d_1(x_1)) - d_2(x_1) \\ &\quad + (d_2(x_1) - d_2(x)) \\ &= (d_1(x) - d_1(x_1)) - a + (d_2(x_1) - d_2(x)). \end{aligned}$$

For each $y_1 \in F_1$, $d(x, y_1) \leq d(x, x_1) + d(x_1, y_1) < \frac{a}{3} + d(x_1, y_1)$. Taking glb,

$$\begin{aligned} d_1(x) &= \text{glb}\{d(x, y_1) \,|\, y_1 \in F_1\} \\ &\leq \frac{a}{3} + \text{glb}\{d(x_1, y_1) \,|\, y_1 \in F\} \\ &= \frac{a}{3} + d_1(x_1), \end{aligned}$$

thus $d_1(x) - d_1(x_1) \leq \frac{a}{3}$.

Similarly, for each point $y_2 \in F_2$

$$d(x_1, y_2) \leq d(x_1, x) + d(x, y_2) < \frac{a}{3} + d(x, y_2).$$

Taking glb, $d_2(x_1) \leq \frac{a}{3} + d_2(x)$, $d_2(x_1) - d_2(x) \leq \frac{a}{3}$.

Thus $d_1(x) - d_2(x) \leq \frac{a}{3} - a + \frac{a}{3} = -\frac{a}{3} < 0$ and $x \in U_1$. Therefore $B(x_1, \frac{a}{3}) \subset U_1$ and U_1 is open.

□

6.1. Exercises

1. Prove that every metric space is T_2.

2. Let X be a first countable space. Then show that X is T_1 iff every convergent seuence has a unique limit.

3. Prove that X is normal iff if F is closed in X and U an open set such that $F \subset U$, then there is an open set V such that $F \subset V \subset \overline{V} \subset U$.

4. Prove that every compact T_2-space is normal.

5. Show the followings:

(1) A subspace of a normal space may not be normal.

(2) A product space of normal spaces need not be normal.

(3) $\prod_{\alpha \in I} X_\alpha$ is T_2 iff X_α is T_2 for all $\alpha \in I$.

6. Let $f, g : X \longrightarrow Y$ be continuous and Y be T_2.
Then prove the following :

(1) $\{x \in X \mid f(x) = g(x)\}$ is closed in X.

(2) If D is dense in X and $f|_D = g|_D$, then $f = g$ on X.

(3) If f is injective, then X is T_2.

6.2 Extensions of Continuous Maps

Recall the followings :

(1) If $(X, \mathcal{T}_1)$ is an indiscrete topological space and $f : X \longrightarrow \mathbb{R}$ is continuous, then f is constant.

(2) If $(X, \mathcal{T}_2)$ is a discrete space, then every map $f : X \longrightarrow \mathbb{R}$ is continuous.

(3) For each $x_1 \neq x_2 \in X$ if there is a continuous map $f : X \longrightarrow \mathbb{R}$ such that $f(x_1) \neq f(x_2)$, then X is T_2.

(4) Denote the complement of A by $X - A := A^c$.

(5) The set of **dyadic numbers** $\{\frac{r}{2^n} \,|\, 0 \leq r \leq 2^n,\ n, r \in \mathbb{Z}\}$ is dense in $[0, 1]$.

Theorem 6.2.1 (Urysohn Lemma). *X is normal iff for each disjoint closed sets A_0 and A_1 in X there is a continuous map $f : X \longrightarrow [0, 1]$ such that $f(A_0) = 0,\ f(A_1) = 1$.*

Proof. ($\Leftarrow$) Let A_0 and A_1 be disjoint closed in X. Then there is a continuous map $f : X \longrightarrow [0, 1]$ such that $f(A_0) = 0,\ f(A_1) = 1$. Let $U_0 = f^{-1}([0, \frac{1}{3}))$, $U_1 = f^{-1}((\frac{2}{3}, 1])$. Then U_0 and U_1 are disjoint open in X and $A_0 \subset U_0$, $A_1 \subset U_1$.
Thus X is normal.

($\Rightarrow$) Suppose that X is normal. Then there are $U_{\frac{1}{2}}$, $V_{\frac{1}{2}}$ disjoint open such that $A_0 \subset U_{\frac{1}{2}}$, $A \subset V_{\frac{1}{2}}$. Then

$$A_0 \subset U_{\frac{1}{2}} \subset X - V_{\frac{1}{2}} \subset X - A_1.$$

For disjoint closed sets A_0 and $X - U_{\frac{1}{2}}$, there are disjoint open sets $U_{\frac{1}{4}}$ and $V_{\frac{1}{4}}$ such that $A_0 \subset U_{\frac{1}{4}}$ and $(X - U_{\frac{1}{2}}) \subset V_{\frac{1}{4}}$. For disjoint closed sets $X - V_{\frac{1}{2}}$ and A_1, there are disjoint open sets $U_{\frac{3}{4}}$ and $V_{\frac{3}{4}}$ such that $(X - V_{\frac{1}{2}}) \subset U_{\frac{3}{4}}$, $A_1 \subset V_{\frac{3}{4}}$. Then

$$A_0 \subset U_{\frac{1}{4}} \subset X - V_{\frac{1}{4}} \subset U_{\frac{1}{2}} \subset X - V_{\frac{1}{2}} \subset U_{\frac{3}{4}} \subset X - V_{\frac{3}{4}} \subset X - A_1.$$

By induction, there are disjoint open sets U_r and V_r, $r = \frac{k}{2^n}$, $k = 1, 2, \cdots, 2^n - 1$, $n = 1, 2, \cdots$ such that

$$A_0 \subset U_{\frac{1}{2^n}} \subset V^c_{\frac{1}{2^n}} \subset U_{\frac{2}{2^n}} \subset V^c_{\frac{2}{2^n}} \subset \cdots \subset U_{\frac{2^n-1}{2^n}} \subset V^c_{\frac{2^n-1}{2^n}} \subset A^c_1.$$

In particular, for a pair of dyadic numbers $r_1 < r_2$,

$$U_{r_1} \subset V^c_{r_1} \subset U_{r_2} \subset V^c_{r_2}.$$

Define a map $f : X \longrightarrow [0, 1]$ by

$$f(x) = \begin{cases} \operatorname{glb}\{r \mid x \in U_r\}, & x \in \cup_r U_r \\ 1, & x \notin \cup_r U_r \end{cases}$$

Then since $A_0 \subset U_r$ for all r, $f(A_0) = 0$, since $U_r \subset A^c_1$ for all r, $f(A_1) = 1$.

Show f is continuous. Let

$$\mathcal{B} := \{[0, a], (c, d), (b, 1], \mid a, b, c, d \text{ are irrationals in } [0, 1]\,\}.$$

Then $\mathcal{B}$ is a basis for the topology of $[0, 1]$.
Since $(c, d) = [0, d) \cap (c, 1]$, $f^{-1}(c, d) = f^{-1}[0, d) \cap f^{-1}(c, 1]$.

Show that $f^{-1}[0, d)$ and $f^{-1}(c, 1]$ are open, for irrational c, $d \in [0, 1]$.

(i) $f^{-1}[0, d) = \cup_{r<d} U_r$:
"$\supset$" If $x \in U_{r_0}$ for some $r_0 < d$, then

$$f(x) \leq r_0 < d, \quad x \in f^{-1}[0, d).$$

"$\subset$" If $x \in f^{-1}[0, d)$, then $f(x) < d$. There is a dyadic number $f(x) < r_0 < d$ such that

$$x \in U_{r_0} \subset \cup_{r<d} U_r.$$

(ii) $f^{-1}(c,1] = \cup_{c<r} V_r$:

" $\supset$ " If $x \in f^{-1}(c,1]$, then $c < f(x)$. There is $c < r_0$ and $x \in U_{r_0}$. There is a dyadic number r_1 such that $c < r_1 < r_0$. Then $V_{r_1}^c \subset U_{r_0}$ and $x \notin V_{r_1}^c$ and $x \in V_{r_1}$. Thus

$$x \in V_{r_1} \subset \cup_{c<r} V_r.$$

" $\subset$ " If $x \in V_r$, for some dyadic $c < r$, then $x \notin V_r^c$ and $x \notin U_r$ since $U_r \subset V_r^c$.

Thus $r < f(x)$ and $c < f(x)$, so $x \in f^{-1}(c,1]$.

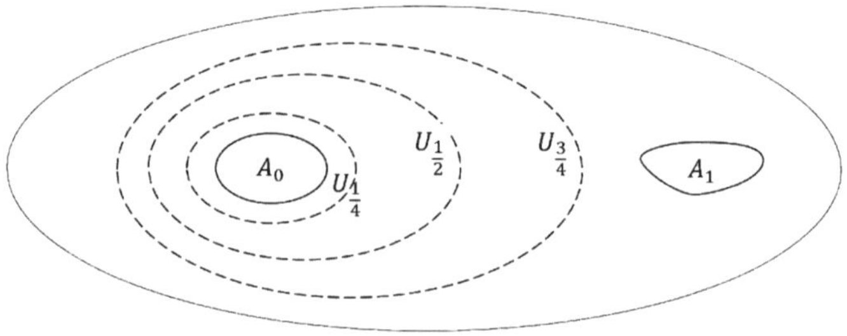

$\square$

Corollary 6.2.2. *Let X be normal and A_0, A_1 disjoint closed subsets in X. Then there is a continuous map $g : X \longrightarrow [a,b]$ such that $g(A_0) = a$, $g(A_1) = b$.*

Proof. By Theorem 6.2.1, there is a continuous map $f : X \longrightarrow [0,1]$ such that $f(A_0) = 0$ and $f(A_1) = 1$. Define a map $h : [0,1] \longrightarrow [a,b]$ by

$$h(t) = (1-t)a + tb$$

for any $t \in [0,1]$. Then h is continuous and $h(0) = a$, $h(1) = b$.
Define $g := h \circ f : X \longrightarrow [0,1] \longrightarrow [a,b]$, then g is a required map. $\square$

Theorem 6.2.3 (Tietz Extension Theorem). *Let X be normal and $X_1 \subset X$ a closed subset. Then any continuous map $f : X_1 \longrightarrow [-1,1]$ has a continuous extension $g : X \longrightarrow [-1,1]$,*
i.e., $g : X \longrightarrow [-1,1]$ is continuous and $g|_{X_1} = f$.

Proof. Let $A_{-1} = f^{-1}[-1, -\frac{1}{3}]$ and $A_1 = f^{-1}[\frac{1}{3}, 1]$. Then A_{-1}, A_1 are closed in X_1 and also closed in X.

By Theorem 6.2.1, there is a continuous map

$$f_1 : X \longrightarrow [-\frac{1}{3}, \frac{1}{3}], \quad f_1(A_{\pm 1}) = \pm\frac{1}{3}.$$

For every $x \in X_1$, $|f(x) - f(x_1)| \leq \frac{2}{3}$. Then

$$f - f_1 : X_1 \longrightarrow [-\frac{2}{3}, \frac{2}{3}]$$

is a continuous map.

Let $A_{-2} = (f - f_1)^{-1}[-\frac{2}{3}, -\frac{2}{3^2}]$ and $A_2 = (f - f_1)^{-1}[\frac{2}{3^2}, \frac{2}{3}]$. Then A_{-2} and A_2 are disjoint closed in X.

By Theorem 6.2.1, there is continuous map

$$f_2 : X \longrightarrow [-\frac{1}{3^2}, \frac{1}{3^2}], \quad f_2(A_{\pm 2}) = \pm\frac{1}{3^2}.$$

For any $x \in X_1$, $|f(x) - f_1(x) - f_2(x)| \leq \frac{4}{9}$.

By induction, for all $n \in \mathbb{N}$, there is a continuous map

$$f_n : X \longrightarrow [-\frac{2^{n-1}}{3^n}, \frac{2^{n-1}}{3^n}], \quad f_n(A_{\pm n}) = \pm\frac{2^{n-1}}{3^n},$$

where

$$A_n = (f - \sum_{i=1}^{n-1} f_i)^{-1}[\frac{2^{n-1}}{3^n}, \frac{2^{n-1}}{3^{n-1}}]$$

and

$$A_{-n} = (f - \sum_{i=1}^{n-1} f_i)^{-1}[-\frac{2^{n-1}}{3^{n-1}}, -\frac{2^{n-1}}{3^n}].$$

Thus for every $x \in X_1$, we have

$$|f(x) - \sum_{i=1}^{n} f_i(x)| \leq (\frac{2}{3})^n.$$

The map $f_n : X \longrightarrow [-1,1]$ and $|f_n(x)| \leq \frac{2^{n-1}}{3^n}$.

For any point $x \in X$,

$$|\sum_{i=1}^{\infty} f_n(x)| \leq \sum_{i=1}^{\infty} |f_n(x)| \leq \sum_{i=1}^{\infty} \frac{2^{n-1}}{3^n} = 1.$$

By Weierstrass M-test, the series $\sum_{n=1}^{\infty} f_n$ converges uniformly to a continuous map $g : X \longrightarrow [-1,1]$, i.e., $g = \sum_{n=1}^{\infty} f_n$.

Moreover, for each $x \in X_1$,

$$|f(x) - g(x)| = \lim_{n\to\infty} |f(x) - \sum_{j=1}^{n} f_j(x)| = \lim_{n\to\infty} (\frac{2}{3})^n = 0.$$

Thus $f = g$ on X_1. □

Remark 6.2.4. (1) **Weierstrass M-test**: Let $f_i : X \longrightarrow \mathbb{R}$ be a continuous map for each i and $s_n = \sum_{i=1}^{n} f_i(x)$. If $|f_i(x)| \leq M_i$, $\forall x \in X$, $\forall i$ and $\sum M_i$ converges, then the sequence s_n converges uniformly to a continuous map s.

(2) Normal $\Longleftrightarrow$ Urysohn $\Longleftrightarrow$ Tietz.

6.2. Exercises

1. Prove that X is normal iff Urysohn Lemma holds. In addition, prove that X is normal iff Tietz Extension Theorem holds.

2. Let C be a closed subset of a normal space X and let $f : C \longrightarrow [a,b]$ be a continuous map. Then show that f has a continuous extension $F : X \longrightarrow [a,b]$.

3. Show that normality and regularity are topological properties.

4. Prove that if X is a normal T_1-space and A is closed in X, then A is also a normal and T_1-space.

6.3 One Point Compactifications

Definition 6.3.1. A space X is **locally compact** if for each point $x \in X$, there is a nbd U of x such that the closure $\bar{U}$ is compact.

Example 6.3.2. (1) $\mathbb{R}$ is locally compact.

(2) $\mathbb{Q}$ is not locally compact.

(3) Let $\{X_i \mid i = 1, 2, \cdots, n\}$ be a family of locally compact spaces. Then $\prod_{i=1}^n X_i$ is locally compact.

Indeed: Let $(x_1, \cdots, x_n) \in \prod_{i=1}^n X_i$. For each $i = 1, 2, \cdots, n$, there is a nbd U_i of x_i in X_i such that $\overline{U_i}$ is compact.

Then $U_1 \times \cdots \times U_n$ is a nbd of $(x_1, \cdots, x_n)$ and

$$\overline{U_1 \times \cdots \times U_n} = \overline{U_1} \times \cdots \times \overline{U_n}$$

is compact in $\prod_{i=1}^n X_i$.

Definition 6.3.3. Let $(X, \mathcal{T}_X)$ be T_1 and $\infty \notin X$. $\widetilde{X} = X \cup \{\infty\}$ is **one point compactification of** X if the topology $\mathcal{T}_{\widetilde{X}}$ of $\widetilde{X}$ has a base $\mathcal{B}_{\widetilde{X}} = \mathcal{T}_X \cup \mathcal{B}$ where

$$\begin{aligned}\mathcal{B} &= \{V \subset \widetilde{X} \mid \widetilde{X} - V : \text{ a compact closed subset of} X\} \\ &= \{\widetilde{X} - F \mid F : \text{ compact closed in } X\}.\end{aligned}$$

Remark 6.3.4. (1) $\mathcal{B}_{\widetilde{X}}$ is a basis for a topology on $\widetilde{X}$.

(2) $\widetilde{X}$ is compact.

(3) $(X, \mathcal{T}_X)$ is a subspace of $(\widetilde{X}, \mathcal{T}_{\widetilde{X}})$.

Example 6.3.5. (1) Let $\widetilde{X} = \mathbb{R} \cup \{\infty\}$ and $X = \mathbb{R}$.

(i) If F is compact in $\mathbb{R}$, then $\widetilde{X} - F$ is a nbd of ∞ in $\widetilde{X}$.

(ii) If $f : S^1 \longrightarrow \widetilde{X}$ is the stereographic projection from $N = (0, 1)$ and $f(N) = \infty$, then f is a homeomorphism.

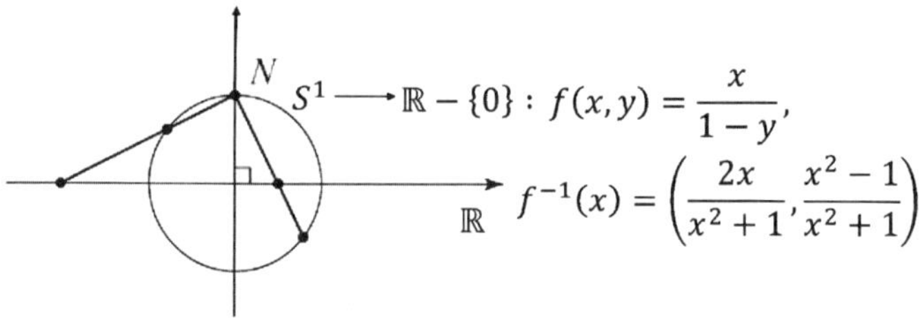

(2) Let $X = \mathbb{R}^2$ and $\widetilde{X} = X \cup \{\infty\}$.

(i) $\widetilde{X} - F$ is a nbd of ∞ if F is compact in X.

(ii) $f : S^2 \longrightarrow \widetilde{X}$ is the stereographic projection from the north pole N and $f(N) = \infty$, that is,

$$f(x_1, x_2, x_3) = \begin{cases} \left(\dfrac{x_1}{1-x_3}, \dfrac{x_2}{1-x_3}, 0\right) & \text{for } \forall (x_1, x_2, x_3) \neq (0,0,1) \\ \\ \infty & \text{for } (x_1, x_2, x_3) = (0,0,1) \end{cases}$$

Then f is a homeomorphism.

(3) As (1) and (2), the one point compactification of $\mathbb{R}^n$: $\widetilde{\mathbb{R}^n} = \mathbb{R}^n \cup \{\infty\}$ and S^n are homeomorphic.

(4) If X is compact, then $\widetilde{X} = X \coprod \{\infty\}$ is a disjoint union and $\{\infty\}$ is closed and open in $\widetilde{X}$.
For example, $\widetilde{S^2} = S^2 \cup \{\infty\}$.

Theorem 6.3.6 (Alexandroff). (1) *If X is T_1, then one point compactificaion $\widetilde{X}$ is T_1 and compact.*

(2) *Let X be a T_1-space. Then X is locally compact and T_2 iff $\widetilde{X}$ is T_2.*

Proof. (1) (i) Let $x_1 \in X$ and $x_2 = \infty \in \widetilde{X}$. Then $U_1 = X$ and $U_2 = \widetilde{X} - \{x_1\}$ are nbds of x_1 and x_2, respectably, and $x_1 \in U_1$, $x_2 \notin U_2$, $x_1 \notin U_2$, $x_2 \in U_2$.
Thus $\widetilde{X}$ is T_1.

(ii) Suppose that $\{U_\alpha \,|\, \alpha \in I\}$ is an open cover of $\widetilde{X}$. Then $\infty \in U_{\alpha_0}$ for some $\alpha_0 \in I$. There is F which is compact closed in X and $\infty \in \widetilde{X} - F \subset U_{\alpha_0}$. Since $\{F \cap U_\alpha \,|\, \alpha \in I\}$ is an open cover of F, there is a finite subcover $\{F \cap U_{\alpha_i} \,|\, i = 1, \cdots, n\}$ of $\{F \cap U_\alpha \,|\, \alpha \in I\}$, i.e.,

$$F = \bigcup_{i=1}^{n} (F \bigcap U_{\alpha_i}) = F \bigcap (\bigcup_{i=1}^{n} U_{\alpha_i}) \subset \bigcup_{i=1}^{n} U_{\alpha_i}.$$

Then

$$\widetilde{X} = F \cup (\widetilde{X} - F) \subset (\cup_{i=1}^{n} U_{\alpha_i}) \cup U_{\alpha_0}.$$

Thus $\widetilde{X}$ is compact.

(2) ($\Leftarrow$) If $\widetilde{X}$ is T_2, then X is T_2. For each point $x \in X$ there are disjoint nbds U_1, U_2 of x, ∞, respectively and $\widetilde{X} - U_2$ compact. Then $U_1 \subset \widetilde{X} - U_2$, $\overline{U_1} \subset \widetilde{X} - U_2$, and $\overline{U_1}$ compact.

($\Rightarrow$) Let $x_1 \in X$ and $x_2 = \infty$. Since X is locally compact, there is a nbd U of x such that $\bar{U}$ is compact. Let $V := \widetilde{X} - \bar{U}$ be open in $\widetilde{X}$. Then $x \in U$, $\infty \in V$ and $U \cap V = \emptyset$. Since X is T_2, $\widetilde{X}$ is T_2.

□

6.3. Exercises

1. Let X be a metric space and $A \subset X$. Then show

(1) A is compact, iff

(2) every infinite subset of A has a limit point in A, iff

(3) every sequence in A contains a subsequence which converges to a point in A.

2. Show that

(1) every discrete space is locally compact.

(2) every indiscrete space is locally compact.

(3) if X is locally compact and A is a closed subset, then A is locally compact.

3. Verify Example 6.3.5 (1)$\sim$(4).

6.4 Paracompact Spaces

Definition 6.4.1. Let X be a space and $\{U_\alpha\}_\alpha$ a family open subsets.

(1) $\{U_\alpha\}$ is an **open cover** of X if $\cup_\alpha U_\alpha = X$.

(2) An open cover $\{V_\beta\}$ is a **refinement** of an open cover $\{U_\alpha\}$ if for each β there is an α such that $V_\beta \subset U_\alpha$.

(3) $\{U_\alpha\}$ is **locally finite** if for each point $x \in X$ there is a nbd W of x such that $W \cap U_\alpha \neq \emptyset$ for only finitely many α.

(4) X is **paracompact** if every open cover of X has a locally finite refinement open cover.

Definition 6.4.2. (1) A space X is **second countable** if the topology of X has a countable basis.

(2) X is **first countable** if each point of X has a locally countable basis.

(3) X is an **n-dimensional manifold** if X is T_2, second countable, and each point of X has a nbd which is homeomorphic to an open subset of $\mathbb{R}^n$, i.e., for each $x \in X$, there are U a nbd of x and V an open subset in $\mathbb{R}^n$ and $h : U \longrightarrow V$ is a homeomorphism.

We introduce some theorems on paracompact spaces without proof.

Theorem 6.4.3. *If X is locally compact, T_2, and second countable, then X is paracompact.*

Example 6.4.4. (1) $\mathbb{R}^n$ is paracompact.

(2) A compact T_2 space is paracompact.

Theorem 6.4.5. *Every paracompact T_2-space is T_4 (normal).*
If a space X is not normal, then X is not paracompact.

Definition 6.4.6. 1 A space $(X, \mathcal{T})$ is **metrizable** if there is a metric d on X such that $\mathcal{T} = \mathcal{T}_d$ the metric topology.

2 X is **locally metrizable** if every point of X has a nbd which is metrizable.

Theorem 6.4.7 (Stone). *Every metrizable space is paracompact.*

Theorem 6.4.8 (Smirnov). *A space X is metrizable iff X is paracompact and locally metrizable.*

Theorem 6.4.9. (1) *Every subspace of a paracompact space is not paracompact.*

(2) *Every product space of a paracompact space is not paracompact.*

(3) *Every closed subset of a paracompact space is paracompact.*

Definition 6.4.10. (1) The **support** of a map $f : X \longrightarrow \mathbb{R}$ is defined by

$$\text{supp}(f) := \overline{f^{-1}(\mathbb{R} - \{0\})}.$$

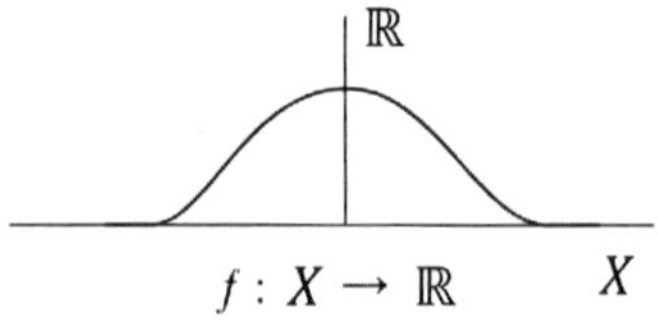

(2) A **partition of unity** of X is a family of maps

$$\{\phi_i : X \longrightarrow \mathbb{R} \,|\, i \in I\}$$

such that

(i) $\{\text{supp}(\phi_i) \mid i \in I\}$ is locally finite,

(ii) for each point $x \in X$, $\phi_i(x) \geq 0$, $\forall i \in I$ and $\sum_{i \in I} \phi_i(x) = 1$.

(3) A partition of unity $\{\phi_i \mid i \in I\}$ is **subordinate to** an open cover $\{U_\alpha \mid \alpha \in J\}$ of X if for each $i \in I$ there is $\alpha \in J$ such that $\text{supp}(\phi_i) \subset U_\alpha$.

Theorem 6.4.11 (Existence of Partition of Unity). *Let X be a paracoompact T_2-space, and $\{U_\alpha \mid \alpha \in J\}$ an open cover of X. Then there is a partition of unity of X subordinate to $\{U_\alpha \mid \alpha \in J\}$.*

Proof. Since X is a paracompact and T_2-space, by Theorem 6.4.7 there are locally finite open coverings $\{W_\alpha\}$ and $\{V_\alpha\}$ of X such that $\overline{W_\alpha} \subset V_\alpha$, $\overline{V_\alpha} \subset U_\alpha$ for all $\alpha \in J$.

Since X is normal, for each α there is continuous map $\psi_\alpha : X \longrightarrow [0,1]$ such that $\psi_\alpha(\overline{W}_\alpha) = 1$, $\psi_\alpha(X - V_\alpha) = 0$. Since ψ_α is nonzero only in V_α, $\psi_\alpha^{-1}(\mathbb{R} - \{0\}) \subset V_\alpha$ and $\text{supp}(\psi_\alpha) \subset \overline{V_\alpha} \subset U_\alpha$.

Since $\{V_\alpha\}$ is locally finite, $\{\overline{V_\alpha}\}$ is also locally finite and $\{\text{supp}(\psi_\alpha)\}$ is locally finite. Define a map $\psi : X \longrightarrow \mathbb{R}$ by $\psi(x) = \sum_\alpha \psi_\alpha(x)$. For each $x \in X$ there is a nbd of x W_x such that $W_x \cap \text{supp}_\alpha \neq \emptyset$ for only finitely many α.
Thus $\psi(x) = \sum_\alpha \psi_\alpha(x)$ is a finite sum, and is continuous.

Now we define a map $\phi_\alpha : X \longrightarrow [0,1]$ by

$$\phi_\alpha(x) = \frac{\psi_\alpha(x)}{\psi(x)}$$

for all α and for all $x \in X$. Then $\{\phi_\alpha\}$ is a desired partition of unity. □

At the beginning of the proof we use following a shrinking Lemma.

Lemma 6.4.12. *Let X be a paracompact T_2-space. Let $\{U_\alpha \mid \alpha \in J\}$ be an open covering of X. Then there is a locally finite open covering $\{V_\alpha \mid \alpha \in J\}$ of X such that $\overline{V_\alpha} \subset U_\alpha$ for each $\alpha \in J$.*

Proof. Since $\{U_\alpha \,|\, \alpha \in J\}$ is an open covering of a paracompact X, there is a locally finite open refine covering $\{W_\beta \,|\, \beta \in I\}$ of X. Define an open covering $\{V_\alpha \,|\, \alpha \in J\}$ by $V_\alpha = \cup\{W_\beta \,|\, W_\beta \subset U_\alpha\}$ for each $\alpha \in J$.
Then $\{V_\alpha \,|\, \alpha \in J\}$ is a locally finite refinement of $\{U_\alpha \,|\, \alpha \in J\}$.
And using the normality of X, we have the lemma. □

Corollary 6.4.13. *Let X be a manifold and U an open subset of X. Let $A = \bar{A} \subset U$. Then a continuous map $\phi : X \longrightarrow \mathbb{R}$ such that*

(1) *for each $x \in X$, $0 \le \phi(x) \le 1$,*

(2) *for every $x \in A$, $\phi(x) = 1$,*

(3) $\mathrm{supp}(\phi) \subset U$.

Proof. Every manifold X is paracompact. The set $\{U, X - A\}$ is an open covering of X. By Theorem 6.4.11, there is a partition of unity $\{\phi, \psi\}$ such that $\mathrm{supp}(\phi) \subset U$, $\mathrm{supp}(\psi) \subset X - A$ and then maps ϕ, $\psi : X \longrightarrow [0,1]$ are continuous, $\phi(x) + \psi(x) = 1$, $\forall x \in X$. If $x \in A$, then $\psi(x) = 0$ and $\phi(x) = 1$. □

Corollary 6.4.14. *If X is a manifold, then there is a continuous proper map $\phi : X \longrightarrow \mathbb{R}$.*

Proof. Let $\{U_\alpha \,|\, \overline{U_\alpha} : \text{compact}, \alpha \in I\}$ be an open cover of X. Then there is a countable partition of unity $\{\phi_i \,|\, i = 1, 2, \cdots, n, \cdots\}$ subordinate to $\{U_\alpha \,|\, \alpha \in I\}$.

Define a map $\phi : X \longrightarrow \mathbb{R}$ by $\phi(x) = \sum_{n=1}^{\infty} n\phi_n(x)$ for every $x \in X$. Then ϕ is a well-defined continuous map. For each $x \in X$ $\phi(x) \in [-k, k]$ for some k, and there is $i \le k$ such that $\phi_i(x) \ne 0$. Then

$$x \in \bigcup_{i=1}^{k} \mathrm{supp}(\phi_i) \subset \bigcup_{i=1}^{k} \overline{U_\alpha},$$

which is compact in X. Since any compact subset of $\mathbb{R}$ is contained in $[-k, k]$ for some k.
Thus ϕ is proper. □

In the proof, we use the existence of countable partition of unity of an open covering of a manifold.

Theorem 6.4.15 (Whitney). *Every n-dimensional manifold is embedded in the $(2n+1)$-dimensional Euclidean space $\mathbb{R}^{2n+1}$.*

6.4. Exercises

1. Let X be paracompact, and $f : X \longrightarrow Y$ a continuous closed surjection. Then show that Y is paracompact.

2. Find an example which a subspace of a paracompact space is not paracompact.

3. Prove that a closed subspace of a paracompact space is paracompact.

4. Find an example which the product of paracompact spaces is not paracompact.

5. If a product is paracompact, then show that each factor is paracompact.

6. Let $\{U_1, \cdots, U_n\}$ be a finite open cover of a T_4-space X. Then show that there exists a partition of unity subordinate to $\{U_1, \cdots, U_n\}$.

7. Let X be a locally compact T_2-space, $F \subset X$ compact, and $F \subset U \subsetneq X$, U open. Then show that there is a continuous map $f : X \longrightarrow [0, 1]$ such that $f(F) = 0$ and $f(X - U) = 1$.

6.5 Classification of Topological Spaces

Definition 6.5.1. A topological space X is **completely regular** (or $\mathbf{T_{3\frac{1}{2}}}$**-space**) if

(1) X is T_1,

(2) for each point $x \in X$ and a closed set $F \subset X - \{x\}$,
there is a continuous map $f : X \longrightarrow [0,1]$ such that $f(x) = 0$ and $f(F) = 1$.

Remark 6.5.2. (1) Every T_4-space is completely regular.

(2) Every completely regular space is T_3.

Indeed, since $U_1 = f^{-1}[0, \frac{1}{3})$, $U_2 = f^{-1}\big((\frac{2}{3}, 1]\big)$ are disjoint open sets and $x \in U_1$, $F \subset U_2$.

Theorem 6.5.3. *Every subset A of a completely regular space X is completely regular.*

Proof. (1) Every subset A of a T_1-space is T_1.

(2) Let $F \subset A$ be closed and $x \in A - F$. Then there is a closed set F_1 in X such that $F = F_1 \cap A$.
There is a continuous map $f : X \longrightarrow [0,1]$ such that $f(x) = 0$, $f(F_1) = 1$. The restriction $f_A : A \longrightarrow [0,1]$ is continuous and $f_A(x) = f(x) = 0$, $f_A(F) = f(F) = f(F_1) = 1$.

□

Theorem 6.5.4. *Every locally compact T_2-space X is completely regular.*

Proof. By Theorem 6.3.6 (2), the one-point compactification $\widetilde{X}$ of a locally compact T_2-space X is a compact T_2-space.
By Theorem 6.1.5, $\widetilde{X}$ is a T_4-space.
By Remark 6.5.2 (1), $\widetilde{X}$ is completely regular.
By Theorem 6.5.3, X is also completely regular. □

Remark 6.5.5. The unit interval $[0,1]$ is compact and T_2.
Let $I_\alpha := I = [0,1]$ for each $\alpha \in A$. Then the product $P = \prod_{\alpha\in A} I_\alpha$ is compact and T_2.
By Theorem 6.5.4, P is completely regular and T_4-space but not a metric space.

Theorem 6.5.6. *Every completely regular space X is a subspace of the product*

$$P = \prod_{\alpha\in A} I_\alpha$$

of the unit intervals.

Proof. Let $A := \{\alpha \,|\, \alpha : X \longrightarrow I \text{ is continuous}\}$ and $I_\alpha := I$, for each $\alpha \in A$. Define a map

$$\varphi : X \longrightarrow P = \prod_{\alpha\in A} I_\alpha$$

by for each $x \in X$ and $\alpha \in A$, $\varphi(x)(\alpha) = \alpha(x)$.

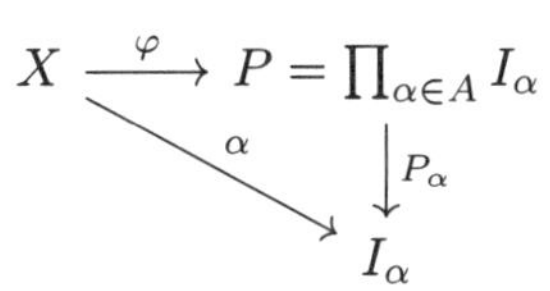

Show $\varphi : X \longrightarrow \varphi(X)$ is a homeomorphism.

(i) For each $\alpha \in A$, and $x \in X$,

$$(P_\alpha \circ \varphi)(x) = P_\alpha(\varphi(x)) = \varphi(x)(\alpha) = \alpha(x).$$

$$\begin{array}{ccc} X & \xrightarrow{\varphi} & P \\ & \searrow^{\alpha} & \downarrow P_\alpha \\ & & I_\alpha \end{array}$$

Then $P_\alpha \circ \varphi = \alpha$ is continuous for each $\alpha \in A$. Thus φ is continuous.

(ii) If $x_0 \neq x_1 \in X$, then by the complete regularity of X there is a continuous map $\alpha : X \longrightarrow [0,1]$, $\alpha(x_0) = 0$, $\alpha(x_1) = 1$. Then $\varphi(x_0)(\alpha) = \alpha(x_0) \neq \alpha(x_1) = \varphi(x_1)(\alpha)$. So, $\varphi(x_0) \neq \varphi(x_1)$.
Thus φ is injective.

(iii) Show $\varphi^{-1} : \varphi(X) \longrightarrow X$ is continuous. Let U be an open subset of X. Show that $\varphi(U)$ is open.

If $U = X$, then $(\varphi^{-1})^{-1}(X) = \varphi(X)$ is open in $\varphi(X)$.

If $U \neq X$, then for any $x \in U$, $X - U$, there is $\alpha \in A$ such that $\alpha(x) = 0$, $\alpha(X - U) = 1$. Then $V := \left(P_\alpha^{-1}[0,1)\right) \cap \varphi(X)$ is open in $\varphi(X)$. Since $P_\alpha(\varphi(x)) = \varphi(x)(\alpha) = \alpha(x) = 0$, $\varphi(x) \in V$. If $\beta \in V$, then there is $y \in X$ such that $\varphi(y) = \beta$ and $P_\alpha(\beta) \in [0,1)$. So,

$$1 > P_\alpha(\beta) = P_\alpha(\varphi(y)) = \varphi(y)(\alpha) = \alpha(y).$$

Then $y \in \alpha^{-1}[0,1) \cap U$, and $\beta = \varphi(y) \subset \varphi(U) \subset \varphi(X)$. If U is open in X, then for each $x \in U$ there is V a nbd of $\varphi(x) \in V \subset \varphi(U) \subset \varphi(X)$, i.e., $\varphi(U)$ is open in $\varphi(X)$.

Hence $\varphi : X \longrightarrow \varphi(X)$ is open map and so $\varphi^{-1} : \varphi(X) \longrightarrow X$ is continuous. □

Definition 6.5.7. (1) A space X is **first countable** if for each $x \in X$ there is a countable family $\{U_n(x) \,|\, U_n(x) \text{ is a nbd of } x, n = 1, \cdots\}$ such that " if $U(x)$ is a nbd of x, then there is n such that $U_n(x) \subset U(x)$".

(2) A space $(X, \mathcal{T})$ is **metrizable** if there is a metric d on X such that $\mathcal{T} = \mathcal{T}_d$.

Note that every metric space is first countable.

Theorem 6.5.8. *Let $I_r := [0,1]$ be the unit interval for $r \in \mathbb{R}$.*

(1) *$P = \prod_{r\in\mathbb{R}} I_r$ is not first countable.*

(2) *The product space $(P, \mathcal{T}_P)$ has no metric d on P such that $\mathcal{T}_P = \mathcal{T}_d$, i.e., $(P, \mathcal{T}_P)$ is not metrizable.*

Proof. (1) Let $x \in P$ such that $x(r) = 0$ for all $r \in \mathbb{R}$, and $\{V_n \,|\, V_n :$ a nbd of $x, n = 1, 2, \cdots\}$ such that for any nbd U of x $V_n \subset U$ for some n, i.e., P is first countable at x.

Each V_n is open in P there is a basic open $\widetilde{V_n}$ such that $x \in \widetilde{V_n} \subset V_n$.

In particular, $x \in \widetilde{V_n} \subset U$ for some basic open $\widetilde{V}_n$. Thus $\{\widetilde{V_n} \,|\, n = 1, 2, \cdots\}$ is a local basis at x. For each n, there is a finite set $F_n \subset \mathbb{R}$ such that

$$\widetilde{V_n} = \prod_{r\in\mathbb{R}} T_{n,r}, \quad T_{n,r} = \begin{cases} I_r, & r \notin F_n \\ U_{n,r}, & r \in F_n \end{cases}$$

where $U_{n,r}$ is open in I_r. Then $C := \cup_{n=1}^{\infty} F_n$ is countable, and $C \subsetneqq \mathbb{R}$.

Choose $r_0 \in \mathbb{R} - C$, then $T_{n,r_0} = I_{r_0}$ for all n. Let $o \in U_{r_0} \subsetneqq I_{r_0}$ and U_{r_0} open. Then

$$\widetilde{P_{r_0}} = U = \prod_{r\in\mathbb{R}} S_r, \quad S_r = \begin{cases} I_r, & r \neq r_0 \\ U_{n_0}, & r = r_0, \end{cases}$$

is an open set P and $x \in U$. However, U does not contain any $\widetilde{V}_n$. This contradicts to the first countability at $x \in P$.

(2) Every metrizable space is first countable.

Indeed, if $(X, \mathcal{T}_d)$ is a metric space and $x \in X$, then

$$\mathcal{B}_x = \{B(x, \frac{1}{n}) \,|\, n \in \mathbb{N}\}$$

is a local basis at x. □

Remark 6.5.9 (Inclusion relations of topological spaces).

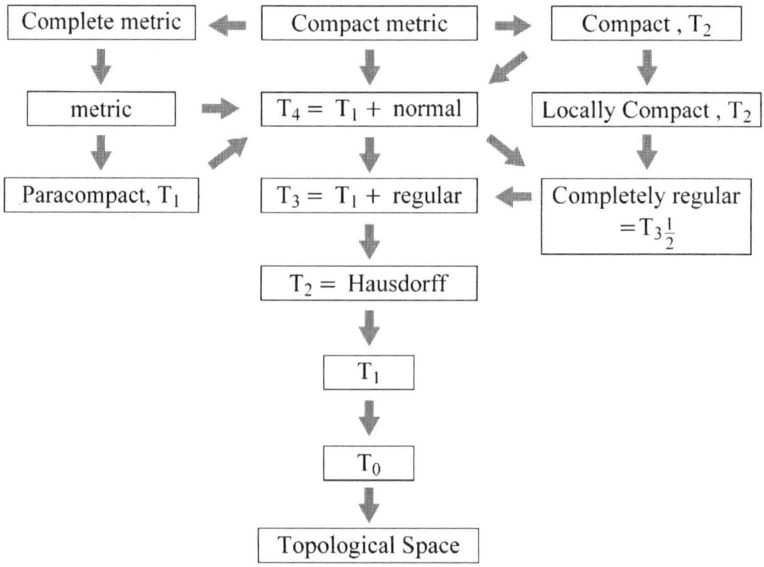

6.5. Exercises

1. Let X be a topological space Then show that the following are equivalent.

(1) X is completely regular.

(2) X is homeomorphic to a subspace of a compact T_2-space.

(3) X is homeomorphic to a subspace of a T_4-space.

2. Prove that every T_3-space with a countable basis is metrizable.

3. Let X be completely regular and

$$c(X,\mathbb{R}) = \{f : X \longrightarrow \mathbb{R} \text{ is continuous}\}.$$

Then show that $c(X, \mathbb{R})$ separates points in X.
i.e., for $x_1 \neq x_2 \in X$, there is a map $f \in c(X, \mathbb{R})$ such that $f(x_1) \neq f(x_2)$.

4. Prove that every subspace of a completely requalr space is completely regular.

Chapter 7

Mapping Spaces

7.1 Complete Metric Spaces

Definition 7.1.1. Let (X, d) be a metric space.

(1) A sequence $< x_n >$ in X **converges to** a point $x \in X$ if

$$\lim_{n \to \infty} d(x_n, x) = 0.$$

(2) A sequence $< x_n >$ is **Cauchy** if $\lim_{n,m \to \infty} d(x_n, x_m) = 0$.

(3) X is a **complete metric space** if every Cauchy sequence in X converges to a point in X.

(4) A map $f : (X_1, d_1) \longrightarrow (X_2, d_2)$ is an **isometric embedding** if for each $x, y \in X_1$

$$d_1(x, y) = d_2(f(x), f(y)).$$

(5) If there is an isometric embedding $f : (X_1, d_1) \longrightarrow (X_2, d_2)$ into a complete metric space X_2, then (X_2, d_2) is called a **completion** of (X_1, d_1).

Remark 7.1.2. (1) If $f : (X_1, d_1) \longrightarrow (X_2, d_2)$ is an isometric embedding, then $f : (X_1, d_1) \longrightarrow (f(X_1), d_2)$ is a homeomorphism.

(2) If $f : (-\frac{\pi}{2}, \frac{\pi}{2}) \longrightarrow \mathbb{R}$, $f(x) = \tan(x)$, then f is a homeomorphism, but $(-\frac{\pi}{2}, \frac{\pi}{2})$ is not complete, whereas $\mathbb{R}$ complete.

Theorem 7.1.3. *For a metric space (X, d), the following are equivalent:*

(1) *X is compact.*

(2) *Every infinite subset A of X has a limit point.*

(3) *Every sequence in X has a convergent subsequence.*

(4) *X is complete and totally bounded*

(i.e., for any $\varepsilon > 0$ there are finite ε-balls $B_\varepsilon(x_1), \cdots, B_\varepsilon(x_n)$ in X such that $X = \cup_{i=1}^n B_\varepsilon(x_i)$).

Proof. (1) $\Longrightarrow$ (2) Let $A \subset X$ such that A infinite, but not has a limit point. Then $A = \bar{A} \subset X$. Since X is compact, A is compact. Since A has no limit point, A has a discrete topology.
Thus A is finite. Contradict!

(2) $\Longrightarrow$ (3) Let $< x_n >$ be a sequence in X. If infinitely many elements in $< x_n >$ are the same, then the infinite elements is a convergent subsequence. If infinitely many elements in $< x_n >$ are distinct and say this set A, then by (2) A has a limit point $x_0 \in X$.

For every k, there exists $x_{n_k} \in B_{x_0}(\frac{1}{k}) \cap A$. Then

$$\lim_{k\to\infty} d(x_{n_k}, x_0) \leq \lim_{k\to\infty} \frac{1}{k} = 0.$$

Thus $\lim_{k\to\infty} x_{n_k} = x_0$.

(3) $\Longrightarrow$ (4) (i) Let $< x_n >$ be Cauchy. By (3) there is a convergent subsequence $< x_{n_k} >$ of $< x_n >$ such that

$$\lim_{k\to\infty} x_{n_k} = x_0 \in X.$$

Since $d(x_n, x_0) \leq d(x_n, x_{n_k}) + d(x_{n_k}, x_0)$, taking n, $n_k \longrightarrow \infty$,

$$\lim_{n\to\infty} d(x_n, x_0) \leq \lim_{n,n_k\to\infty} d(x_n, x_{n_k}) + \lim_{n_k\to\infty} d(x_{n_k}, x_0) = 0.$$

Thus, X is complete.

(ii) Show that X is totally bounded. If not, then there is $\varepsilon > 0$, any finite ε-balls can not cover X. Choose a point $x_1 \in X$. Then there is $x_2 \in X - B_\varepsilon(x_1)$, and $d(x_1, x_2) \geq \varepsilon$. Thus

$$x_n \in X - \bigcup_{k=1}^{n-1} B_\varepsilon(x_k).$$

We have a sequence $< x_n >$ such that $d(x_i, x_j) \geq \varepsilon$, for all i, j.
By (3), there is a subsequence $< x_{n_k} >$ of $< x_n >$ such that

$$\lim_{k\to\infty} x_{n_k} = x_0 \in X \quad \text{for some point } x_0.$$

There is k_0 such that $x_{n_k} \in B_{x_0}(\frac{\varepsilon}{2})$ for all $k > k_0$.
Thus for k_1, $k_2 > k_0$,

$$d(x_{n_{k_1}}, x_{n_{k_2}}) \leq d(x_{n_{k_1}}, x_0) + d(x_0, x_{n_{k_2}}) < \varepsilon.$$

This contradicts to our construction.

(4) $\Longrightarrow$ (1) (i) (4) $\Longrightarrow$ (3) Let $< x_n >$ be a sequence in X. For $1 > 0$, by (4), there are finite balls $B_1(1), \cdots, B_{k_1}(1)$ which cover X. There is j_1 such that $1 \leq j_1 \leq k_1$ and an infinite set $J_1 \subset \mathbb{N}$ such that $x_n \in B_{j_1}(1)$ for all $n \in J_1$.

Let n_1 be the first $x_{n_1} \in B_{j_1}(1)$. For $\frac{1}{2} > 0$, by (4), there are finitely many balls $B_1(\frac{1}{2}), \cdots, B_{k_2}(\frac{1}{2})$ which cover X and there is j_2 such that $1 \leq j_2 \leq k_2$ and an infinite set $J_2 \subset J_1$ such that $x_n \in B_{j_2}(\frac{1}{2})$ for all $n \in J_2$.

Let n_2 be the first $x_{n_2} \in B_{j_2}(\frac{1}{2})$, $n_2 > n_1$. By the induction we have a sequence $< x_{n_k} >$. If $\ell \geq k$, then $x_{n_\ell} \in B_{j_k}(\frac{1}{k})$. If $\ell, m \geq k$, then $d(x_{n_\ell}, x_{n_m}) < \frac{2}{k}$.
Thus the sequence $< x_{n_k} >$ is Cauchy.
Since X is complete, $< x_{n_k} >$ converges in X.

(ii) $(3),(4) \Longrightarrow (1)$ Let $\mathcal{U}$ be an open cover of X. Show that there is $n \in \mathbb{N}$ such that every $\frac{1}{n}$-ball is contained in an element of $\mathcal{U}$.

Suppose that for each $n \in \mathbb{N}$, there exists $B_{x_n}(\frac{1}{n})$ is not contained in any element of $\mathcal{U}$.
By (3) the sequence $< x_n >$ has a convergent subsequence $< x_{n_k} >$,

$$\lim_{k\to\infty} x_{n_k} := x_0 \in U_0 \in \mathcal{U},$$

for some $x_0 \in X$, and $U_0 \in \mathcal{U}$.

Since U_0 is open, there exists $r > 0$ such that $B_{x_0}(r) \subset U_0$. Since $\lim_{k\to\infty} x_{n_k} = x_0$, there exists k_0 such that $B_{x_k} \in B_{x_0}(\frac{r}{2})$ for any $k > k_0$. If $\frac{1}{n_\ell} < \frac{r}{2}$ and $\ell \geq k_0$, then $B_{x_{n_k}}(\frac{1}{n_\ell}) \subset B_{x_0}(r) \subset U_0$.
This contradicts to our construction of $< x_n >$.
Since X is totally bounded, we prove (1). □

Theorem 7.1.4 (Lebesgue Number Theorem). *Let X be a compact metric space, and $\mathcal{U}$ an open cover of X. Then there is $\varepsilon > 0$ such that every ε-ball is contained in some element in $\mathcal{U}$. In this case, the ε is called* **Lebesgue number** *of $\mathcal{U}$.*

7.1. Exercises

1. Prove that every closed subspace of a complete metric space is complete.

2. Prove that a metric space X is complete iff every infinite totaly bounded subset of X has a limit point.

3. Show that a metric space X is totally bounded iff every sequence in X contains a Cauchy subsequence.

4. Let $(X, \mathcal{T}_d)$ be a metric space. Prove the followings :

(1) If $\lim_{n\to\infty} a_n = a$ and $\lim_{n\to\infty} b_n = b$ in X, then show

$$\lim_{n\to\infty} d(a_n, b_n) = d(a, b).$$

(2) Let

$$\bar{d}(a, b) = \frac{d(a, b)}{1 + d(a, b)}$$

for each a, $b \in X$. Then $\bar{d}$ is a metric on X and $\mathcal{T}_d = \mathcal{T}_{\bar{d}}$.

5. Prove that Euclidean space $(\mathbb{R}^n, d)$ is complete.

6. Let (X, d) be a metric space. Let

$$c[X] = \{< a_n >: \text{ a Cauchy sequence in } X\}$$

and define

$$< a_n > \sim < b_n > \quad \text{iff} \quad \lim_{n\to\infty} d(a_n, b_n) = 0.$$

(1) Show $\sim$ is an equivalence relation on $c[X]$. Denote $c[X]/\sim$ by $\widetilde{X}$. Define

$$\widetilde{d}([< a_n >], [< b_n >]) = \lim_{n\to\infty} d(a_n, b_n),$$

where $[< a_n >]$, $[< b_n >] \in \widetilde{X}$.

(2) Show $\widetilde{d}$ is a metric on $\widetilde{X}$.

(3) Show $h : X \longrightarrow \widetilde{X}$ given by $h(a) = [< a, a, \cdots >]$ is an isometric embedding, and $\overline{h(X)} = \widetilde{X}$.

(4) $(\widetilde{X}, \widetilde{d})$ is complete.

7.2 Category Theorem

Definition 7.2.1. Let X be a topological space and $A \subset X$.

(1) A is **nowhere dense** if $\bar{A}$ does not contain any open set in X, i.e., $X - \bar{A}$ is dense.

(2) A is **first category** if A is a countably many union of nowhere dense subsets.

(3) A is **second category** if A is not first category.

Theorem 7.2.2. *Let (X, d) be a complete metric space. If U_n is an open dense subset of X for each n, then*

$$\bigcap_{n=1}^{\infty} U_n \neq \emptyset.$$

Proof. Let $x_1 \in U_1$. Then there is a ball $B_{x_1}(r_1) \subset U_1 \subset \overline{U_1} = X$. Since U_i is dense, $B_{x_1}(r_1) \cap U_2 \neq \emptyset$. Choose $x_2 \in B_{x_1}(r_1) \cap U_2$, $x_1 \neq x_2$.

Choose r_2 such that

$$0 < r_2 < \min\{\frac{r_1}{2}, d(x_1, x_2)\},$$

and $B_{x_2}(r_2) \subset B_{x_1}(r_1) \cap U_2$. If $x \in \overline{B_{x_2}(r_2)}$, then

$$d(x, x_1) \leq d(x, x_2) + d(x_2, x_1) \leq r_2 + d(x_1, x_2) < r_1.$$

Thus $x \in B_{x_1}(r_1)$ and so $\overline{B_{x_2}(r_2)} \in B_{x_1}(r_1)$.

By induction, we have a sequence $< x_n >$ in X such that

$$x_n \in B_{x_n}(r_n), \quad r_n < \frac{r_1}{2^{n-1}}, \quad \overline{B_{x_{n+1}}(r_{n+1})} \subset U_n.$$

For any $\varepsilon > 0$, choose $r_{n_0} < \frac{\varepsilon}{2}$.

If $n > n_0$, then $x_n \in B_{x_n}(r_n)$. For m, $n > 0$,

$$d(x_m, x_n) \leq d(x_n, x_{n_0}) + d(x_{n_0}, x_n) < r_{n_0} + r_{n_0} < \varepsilon.$$

Thus the sequence $< x_n >$ is Cauchy.

Since X is complete, $\lim_{n\to\infty} x_n = x_0$ for some $x_0 \in X$. For each n,

$$x_0 = \lim_{k\to\infty} x_{n+k} \in \overline{B_{x_{n+1}}(r_{n+1})} \subset B_{x_n}(r_n) \subset U_n.$$

Thus $x_0 \in \cap_{n=1}^{\infty} U_n$. □

Theorem 7.2.3 (Baire Category Theorem). *Every complete metric space (X, d) is second category.*

Proof. If $X = \cup_{n=1}^{\infty} N_n$, where N_n is nowhere dense in X for $\forall n$, then $X = \cup_{n=1}^{\infty} \overline{N_n}$, and

$$X - \cup_{n=1}^{\infty} \overline{N_n} = \cap_{n=1}^{\infty} (X - \overline{N_n}) = \emptyset.$$

While $X - \overline{N_n}$ is an open dense subset of X for each n.
By Theorem 7.2.2,

$$\bigcap_{i=1}^{\infty} (X - \overline{N_n}) \neq \emptyset.$$

Thus we have a contradiction. □

7.2. Exercises

1. Prove that every complete metric space is second category.

2. Show that a countable union of first category spaces is of finite category.

3. Prove that the intersection of any countable family of open dense sets in a locally compact space X is dense.

7.3 Mapping Spaces

Definition 7.3.1. $(N, ||\cdot||)$ is a **normed linear space** if

(1) N is a linear space (i.e., a vector space over $\mathbb{R}$).

(2) A norm $||\cdot|| : N \longrightarrow \mathbb{R}$ is a map such that for each a, $b \in N$, $\lambda \in \mathbb{R}$

(i) $||a|| \geq 0, \quad ||a|| = 0$ iff $a = 0$,

(ii) $||\lambda a|| = |\lambda|\, ||a||$,

(iii) $||a + b|| \leq ||a|| + ||b||$.

Remark 7.3.2. A normed linear space $(N, ||\cdot||)$ is a metric space with metric

$$d : N \times N \longrightarrow \mathbb{R}, \quad d(a, b) = ||a - b||.$$

Indeed,

(1) $d(a, b) = ||a - b|| \geq 0$.

(2) $d(a, b) = 0$ iff $||a - b|| = 0$ iff $a = b$.

(3) $d(a, b) = ||a - b|| = ||(-1)(b - a)|| = |-1|\, ||b - a|| = d(b, a)$.

(4) $d(a, c) = ||a-c|| = ||(a-b)+(b-c)|| \leq ||a-b||+||b-c|| = d(a, b)+d(b, c)$.

Thus (N, d) is a metric space. In this case $||\cdot||$ is called a **norm** on N.

Definition 7.3.3. A normal linear space $(N, ||\cdot||)$ is a **Banach space** if (N, d) is a complete metric space.

Example 7.3.4. Let

$$N := C([0, 1]) = \{f : [0, 1] \longrightarrow \mathbb{R} \text{ is a continuous map}\}.$$

Define a norm $||\cdot|| : N \longrightarrow \mathbb{R}$ by $||f|| = \max_{x \in [0,1]} |f(x)|, \quad \forall f \in N$. Then $(N, ||\cdot||)$ is a normal linear space.

Theorem 7.3.5. *Let $\{f_n : (X, d) \longrightarrow \mathbb{R}\}$ be a sequence $< f_n >$ of continuous maps. If $< f_n >$ is uniformly convergent to a map $f : X \longrightarrow \mathbb{R}$, then f is also continuous.*

Remark 7.3.6. In Example 7.3.4, by Theorem 7.3.5, the $(N, ||\cdot||)$ is a Banach space.

Theorem 7.3.7. *There is a continuous map $f \in C([0,1])$ which is nowhere differentiable.*

For example, a continuous sawtooth map $f : [0,1] \longrightarrow \mathbb{R}$.

7.3. Exercises

1. For each element $(a_1, \cdots, a_n) \in \mathbb{R}^n$, we define a map

$$||\cdot|| : \mathbb{R}^n \longrightarrow \mathbb{R}$$

by

$$||(a_1, \cdots, a_n)|| = \Big(\sum_{i=1}^{n} a_i^2\Big)^{1/2}.$$

Show (1) $||\cdot||$ is a norm on $\mathbb{R}^n$.

(2) For any points $\mathbf{a} = (a_1, \cdots, a_n)$, $\mathbf{b} = (b_1, \cdots, b_n) \in \mathbb{R}^n$, show

(i) $\sum_{i=1}^{n} |a_i b_i| \leq ||\mathbf{a}|| \cdot ||\mathbf{b}||$,

(ii) $||\mathbf{a} + \mathbf{b}|| \leq ||\mathbf{a}|| + ||\mathbf{b}||$.

2. Let $C([0,1]) = \{f : [0,1] \longrightarrow \mathbb{R}$ is continuous$\}$, and its norm

$$||f|| := \max\{|f(x)| \,|\, x \in [0,1] \text{ for each } \ f \in c([0,1])\}.$$

Then show that $\Big(C([0,1]), ||\cdot||\Big)$ is a normal linear space.

3. Let $\mathbb{R}^\infty$ is defined by

$$\mathbb{R}^\infty = \{(a_n) = (a_1, \cdots, a_n, \cdots) \,|\, a_i \in \mathbb{R},\ i = 1, \cdots, \sum_{i=1}^{\infty} a_i^2 < \infty\}.$$

Prove (1) Let $||(a_n)|| = (\sum_{i=1}^{n} a_i^2)^{1/2}$ for each $(a_n) \in \mathbb{R}^\infty$.
Then, $(\mathbb{R}^\infty, ||\cdot||)$ is a normed linear space.

(2) Let $d((a_n),(b_n)) = ||(a_n) - (b_n)||$ for each (a_n), $(b_n) \in \mathbb{R}^\infty$.
Then $(\mathbb{R}^\infty, d)$ is a metric space.

Chapter 8

Fundamental Groups

In this chapter, we assume that every X is a topological space, every map is continuous, and $I = [0, 1]$ is the unit interval.

8.1 Homotopy of Paths

Definition 8.1.1. (1) A map $\alpha : I \longrightarrow X$ is called a **path** in X.

(2) Two paths $\alpha, \beta : I \to X$ is **homotopic with end points held fixed**, if there is a map $H : I \times I \longrightarrow X$ such that

(i) $H(s, 0) = \alpha(s)$, $s \in I$

(ii) $H(s, 1) = \beta(s)$, $s \in I$

(iii) $H(0, t) = \alpha(0) = \beta(0)$, $t \in I$

(iv) $H(1, t) = \alpha(1) = \beta(1)$, $t \in I$.

In this case, H is called a **homotopy** from α to β.
Denote it by $H : \alpha \simeq \beta \quad \text{rel } (0, 1)$.

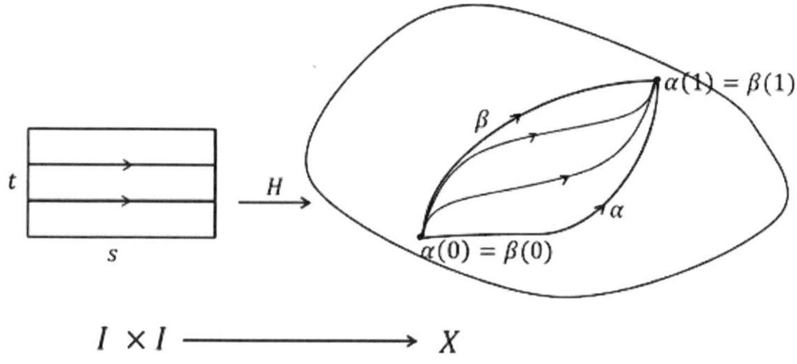

Theorem 8.1.2. *The homotopy* $H : \alpha \simeq \beta \quad \text{rel } (0,1)$ *is an equivalence relation, i.e.,*

(1) $\alpha \simeq \alpha \quad \text{rel } (0,1)$ *for all* α,

(2) *If* $\alpha \simeq \beta \quad \text{rel } (0,1)$, *then* $\beta \simeq \alpha \quad \text{rel } (0,1)$,

(3) *If* $\alpha \simeq \beta \quad \text{rel } (0,1)$ *and* $\beta \simeq \gamma \quad \text{rel } (0,1)$, *then* $\alpha \simeq \gamma \quad \text{rel } (0,1)$.

Proof. (1) Let $H : I \times I \longrightarrow X$, $H(s,t) = \alpha(s)$, $\forall s \ \forall t \in I$. Then H is continuous and $H : \alpha \simeq \alpha \quad \text{rel } (0,1)$.

(2) If $H : I \times I \longrightarrow X$, $H : \alpha \simeq \beta \quad \text{rel } (0,1)$,
then $H' : I \times I \longrightarrow X$ given by
$H'(s,t) = H(1-s,t)$ is $H' : \beta \simeq \alpha \quad \text{rel } (0,1)$.

(3) If $H_1 : \alpha \simeq \beta \quad \text{rel } (0,1)$ and $H_2 : \beta \simeq \gamma \quad \text{rel } (0,1)$, then the map $H_3 : I \times I \longrightarrow X$ given by

$$H_3(s,t) = \begin{cases} H_1(s,2t), & s \in I, \ \ 0 \le t \le \dfrac{1}{2} \\ H_1(s,2t-1), & s \in I, \ \ \dfrac{1}{2} \le t \le 1, \end{cases}$$

is $H_3 : \alpha \simeq \gamma \quad \text{rel } (0,1)$.

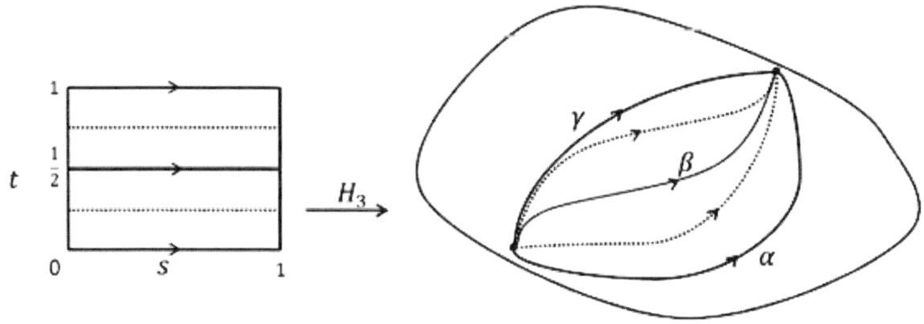

□

The set $[\alpha] := \{\beta | \alpha \simeq \beta \quad \text{rel } (0,1)\}$ is called the **homotopy class** of α.

Definition 8.1.3. Let $\alpha, \beta : I \longrightarrow X$ be paths and $\alpha(1) = \beta(0)$. Then the path $\alpha \cdot \beta : I \longrightarrow X$ is given by

$$\alpha \cdot \beta(t) = \begin{cases} \alpha(2t), 0 \le t \le \dfrac{1}{2} \\ \beta(2t-1), \dfrac{1}{2} \le t \le 1 \end{cases}$$

is called the **product** of α and β.

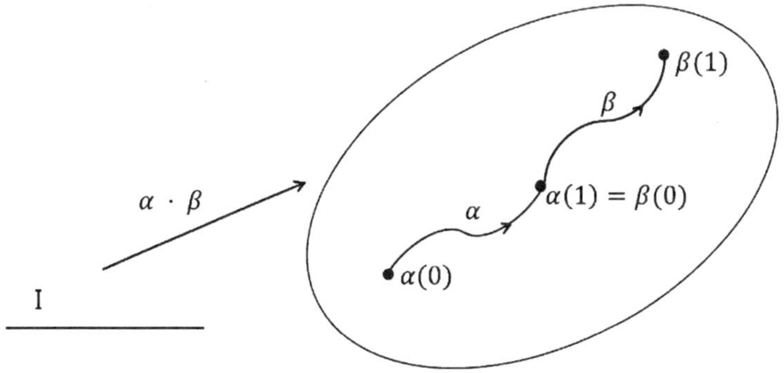

Theorem 8.1.4. *If $H_1 : \alpha \simeq \alpha \quad \text{rel } (0,1)$, $H_2 : \beta \simeq \beta \quad \text{rel } (0,1)$ and $\alpha(1) = \beta(0)$, then $\alpha \cdot \beta \simeq \alpha' \cdot \beta' \quad \text{rel } (0,1)$.*

Proof. Define a homotopy

$$H : I \times I \longrightarrow X \quad \text{by}$$

$$H(s,t) = \begin{cases} H_1(2s,t), & 0 \le s \le \frac{1}{2}, \quad 0 \le t \le 1 \\ H_2(2s-1,t), & \frac{1}{2} \le s \le 1, \quad 0 \le t \le 1. \end{cases}$$

Then

$$H(s,0) = \left\{ \begin{matrix} H_1(2s,0), & 0 \le s \le \frac{1}{2} \\ H_2(2s-1,0), & \frac{1}{2} \le s \le 1 \end{matrix} \right\}$$

$$= \left\{ \begin{matrix} \alpha(s) \\ \beta(2s-1) \end{matrix} \right\} = \alpha \cdot \beta(s).$$

$$H(s,1) = \left\{ \begin{matrix} H_1(2s,1), & 0 \le s \le \frac{1}{2} \\ H_2(2s-1,1), & \frac{1}{2} \le s \le 1 \end{matrix} \right\}$$

$$= \left\{ \begin{matrix} \alpha'(s) \\ \beta'(2s-1) \end{matrix} \right\} = \alpha' \cdot \beta'(s).$$

Also $H(0,t) = H_1(0,t) = \alpha(0), \quad 0 \le t \le 1$

$H(1,t) = H_2(1,t) = \beta(0), \quad 0 \le t \le 1.$

Thus $H : \alpha \cdot \beta \simeq \alpha' \cdot \beta'$ rel $(0,1)$.

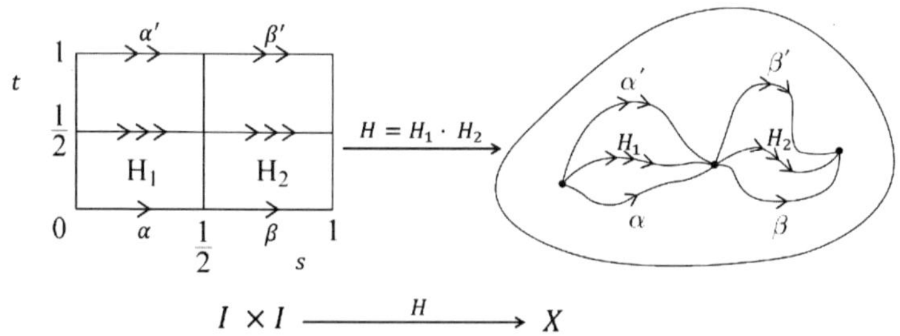

□

8.2 Fundamental Groups

A path $\alpha : I \longrightarrow X$ with $\alpha(0) = \alpha(1) = x_0$ is called a **loop** at x_0 in X.

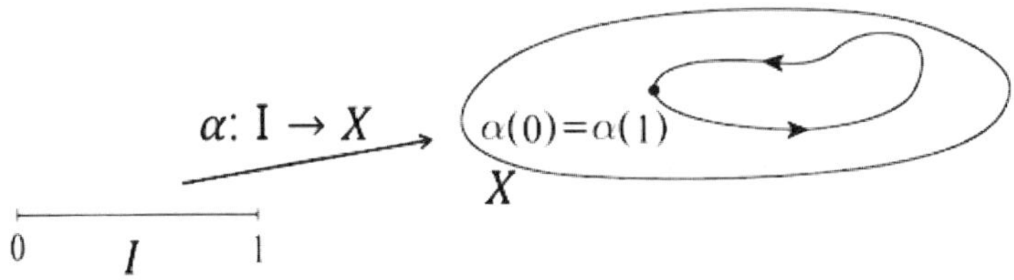

Theorem and Definition 8.2.1. *For $x_0 \in X$, let $\pi_1(X, x_0) = \{[\alpha] \mid \alpha : \text{a loop at } x_0 \in X\}$, and $x_0 : I \longrightarrow X$ the constant loop, and $\alpha^{-1} : I \longrightarrow X$ defined by $\alpha^{-1}(t) = \alpha(1-t)$, $\forall t \in I$, and define a binary operation on $\pi_1(X, x_0)$ by $[\alpha] \cdot [\beta] = [\alpha \cdot \beta]$ for all $[\alpha], [\beta] \in \pi_1(X, x_0)$. Then $[x_0]$ is the identity in $\pi_1(X, x_0)$ and $[\alpha^{-1}]$ is the inverse of $[\alpha]$ for all $[\alpha] \in \pi_1(X, x_0)$, and $\pi_1(X, x_0)$ is a group under the operation. $\pi_1(X, x_0)$ is called the* **fundamental group** *of X at x_0.*

Proof. (1) $[x_0]$ is the identity in $\pi_1(X, x_0)$:

Show $\alpha \cdot x_0 \simeq \alpha \quad \text{rel } (0,1)$ $H : I \times I \longrightarrow X$ defined by

$$H(s,t) = \begin{cases} \alpha\left(\dfrac{2s}{t+1}\right), & 0 \le s \le \dfrac{t+1}{2}, \\ x_0, & \dfrac{1}{2} \le s \le 1. \end{cases}$$

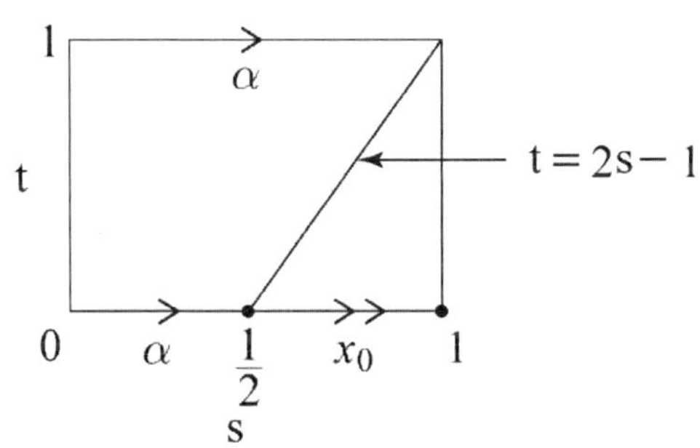

If $s = \frac{t+1}{2}$, then $\alpha(\frac{2s}{t+1}) = \alpha(1) = x_0$.
Thus H is continuous.

$$H(s,0) = \begin{Bmatrix} \alpha(2s) & 0 \le s \le \dfrac{1}{2} \\ x_0 & \dfrac{1}{2} \le s \le 1 \end{Bmatrix} = \alpha \cdot x_0(s) \quad \forall s \in I,$$

$$H(s,1) = \begin{Bmatrix} \alpha(s) & 0 \le s \le 1 \\ x_0 & s = 1 \end{Bmatrix} = \alpha(s) \quad \forall s \in I.$$

$H(0,t) = \alpha(0) = x_0,$

$$H(1,t) = \begin{Bmatrix} \alpha(\dfrac{2}{t+1}) & 1 \le \dfrac{t+1}{2} \\ x_0 & \dfrac{t+1}{2} \le 1 \end{Bmatrix} = \alpha(1) = x_0.$$

Thus $\alpha \cdot x_0 \simeq \alpha \quad \text{rel } (0,1)$ and $[\alpha] \cdot [x_0] = [\alpha \cdot x_0] = [\alpha]$.

(2) $[\alpha][\alpha^{-1}] = [x_0]$:
Show $x_0 \simeq \alpha \cdot \alpha^{-1} \quad \text{rel } (0,1)$.
Define $I \times I \longrightarrow X$ by

$$H(s,t) = \begin{cases} \alpha(\dfrac{2s}{t+1}), & 0 \le 2s \le t \\ \alpha(t), & t \le 2s \le 2-t \\ \alpha^{-1}(2s-1), & 2-t \le 2s \le 2. \end{cases}$$

Then H is continuous and $H : \alpha \cdot \alpha^{-1} \simeq x_0 \quad \text{rel } (0,1)$.

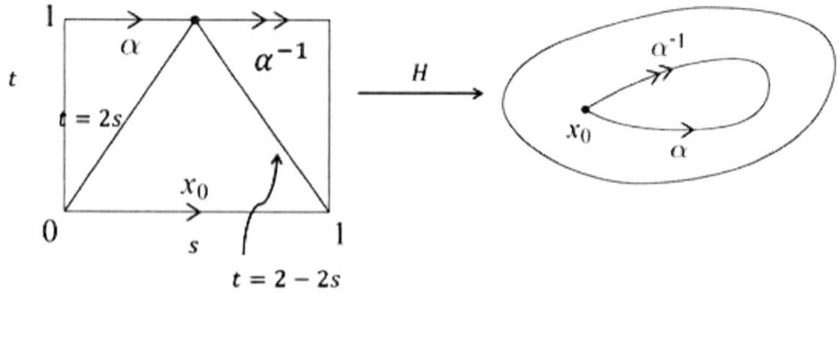

Thus $[\alpha][\alpha^{-1}] = [\alpha][\alpha^{-1}] = [\alpha \cdot \alpha^{-1}] = [x_0]$.

(3) $([\alpha][\beta])[\gamma] = [\alpha]([\beta][\gamma])$ for every $[\alpha], [\beta], [\gamma] \in \pi_1(X, x_0)$.
Show that $(\alpha \cdot \beta) \cdot \gamma \simeq \alpha \cdot (\beta \cdot \gamma) \quad \text{rel } (0, 1)$.

Define $H : I \times I \longrightarrow X$ by

$$H(s,t) = \begin{cases} \alpha(\dfrac{4s}{t+1}), & 0 \le s \le \dfrac{1}{4}(t+1), \\ \beta(4s - t - 1), & \dfrac{t+1}{4} \le s \le \dfrac{1}{4}(t+2), \\ \gamma(\dfrac{4s-t-2}{2-t}), & \dfrac{1}{4}(t+2) \le s \le 1. \end{cases}$$

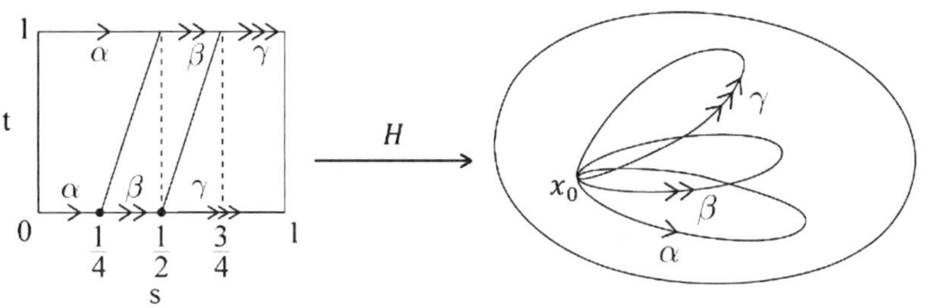

Then H is continuous and $H : (\alpha \cdot \beta) \cdot \gamma \simeq \alpha \cdot (\beta \cdot \gamma) \quad \text{rel } (0, 1)$.

So, $([\alpha][\beta])[\gamma] = [\alpha \cdot \beta][\gamma] = [(\alpha \cdot \beta) \cdot \gamma]$

$$= [\alpha \cdot (\beta \cdot \gamma)] = [\alpha][\beta \cdot \gamma] = [\alpha]([\beta][\gamma]).$$

Thus $\pi_1(X, x_0)$ is a group.

□

Theorem 8.2.2. *Let $\sigma : I \longrightarrow X$ be a path with $\sigma(0) = x_0$ and $\sigma(1) = x_1$. Then the map $\sigma_\sharp : \pi_1(X, x_0) \longrightarrow \pi_1(X, x_1)$ defined by $\sigma_\sharp([\alpha]) = [\sigma^{-1}\alpha\sigma]$ is an isomorphism.*

Proof. For every $[\alpha_1], [\alpha_2] \in \pi_1(X, x_0)$,

$$\sigma_\sharp([\alpha_1][\alpha_2]) = \sigma_\sharp([\alpha_1 \cdot \alpha_2]) = [\sigma^{-1} \cdot (\alpha_1 \cdot \alpha_2) \cdot \sigma]$$
$$[\sigma^{-1} \cdot \alpha_1 \cdot \sigma][\sigma^{-1} \cdot \alpha_2 \cdot \sigma] = \sigma_\sharp([\alpha_1]) \cdot \sigma_\sharp([\alpha_2])$$

Thus $\sigma_\sharp$ is a homomorphism.

Moreover, the homomorphism $\sigma_\sharp$ has its inverse

$$(\sigma^{-1})_\sharp : \pi_1(X, x_1) \longrightarrow \pi_1(X, x_0).$$

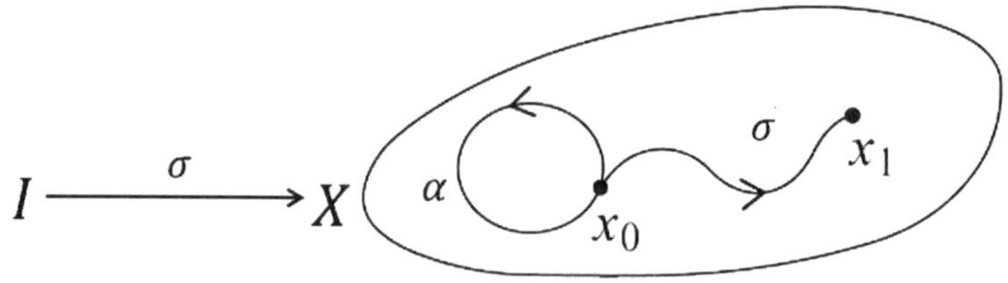

Therefore $\sigma_\sharp$ is an isomorphism. □

If X is path connected, i.e., any two points in X can be joined by a path, then all fundamental groups of X are isomorphic.

Theorem 8.2.3. *For a map $f : (X, x_0) \longrightarrow (Y, y_0)$ with $f(x_0) = y_0$, we define $f_* : \pi_1(X, x_0) \longrightarrow (Y, y_0)$ by $f_*([\alpha]) = [f \cdot \alpha]$. Then*

(1) *f_* is a homomorphism.*

(2) *$(I_X)_* = I_{\pi(X,x_0)} : \pi_1(X, x_0) \longrightarrow \pi_1(X, x_0)$ is the identity.*

(3) *If $g : (Y, y_0) \longrightarrow (Z, z_0)$, then $(g \circ f)_* = g_* \circ f_*$.*

Proof. Proof is clear by definition of f_*. □

We have a correspondence

$$\pi : \text{Category of topological spaces } \{(X, x_0) \xrightarrow{f} (Y, y_0)\}$$
$$\longrightarrow \text{Category of groups } \{\pi_1(X, x_0) \xrightarrow{f_*} \pi_1(Y, y_0)\}.$$

Then by Theorem 8.2.3, π is a covariant functor.

Definition 8.2.4. (1) X is **simply connected** if X is path connected and its fundamental group $\pi_1(X) = \{1\}$.

(2) Two maps $f, g : X \longrightarrow Y$ are **homotopic** if there is homotopy $H : X \times I \longrightarrow Y$ such that $H(x, 0) = f(x)$ and $H(x, 1) = g(x)$ $\forall x \in X$. Denote it by $f \simeq g$.

(3) X and Y are **homotopic** or **homotopy equivalence** if there are maps $f : X \longrightarrow Y$ and $g : Y \longrightarrow X$ such that $g \circ f \simeq I_X$, $f \circ g \simeq I_Y$.

(4) If X is homotopic to a point, then we say that X is **contractable**.

If X is contractable, then X is simply connected.

Theorem 8.2.5. *For maps* $f,\, g : (X, x_0) \longrightarrow (Y, y_0)$ *if*

$$H : (X, x_0) \times I \longrightarrow (Y, y_0),$$

$f(x) = H(x, 0)$, $g(x) = H(x, 1)$ *and* $H(x_0, t) = f(x_0) = g(x_0) = y_0$ $\forall t \in I$, *then* $f_* = g_* : \pi_1(X, x_0) \longrightarrow \pi(Y, y_0)$.

Proof. For each $[\alpha] \in \pi_1(X, x_0)$, the composite

$$\sum \times I \xrightarrow{\alpha \times I} X \times I \longrightarrow []HY$$

say $f \circ \alpha \simeq g \circ \alpha$ rel $(0, 1)$. Then $f_*([\alpha]) = [f \circ \alpha] = [g \circ \alpha] = g_*([\alpha])$. Thus $f_* = g_*$. □

8.3 Fundamental Groups of Circle

Let $S^1 = \{e^{2\pi it} | 0 \leq t \leq 1\}$ be the unit circle. The map

$$p : (\mathbb{R}, +) \longrightarrow (S^1, \cdot)$$

given by $p(t) = e^{2\pi it}$, is a homomorphism, and $p([0,1]) = p([-\frac{1}{2}, \frac{1}{2}]) = S^1$, the restriction $p : (-\frac{1}{2}, \frac{1}{2}) \longrightarrow S^1 - \{(-1,0)\}$ is homeomorphism. Moreover $p^{-1}(1) = \mathbb{Z}$. Also p is a local homeomorphism.

$$\begin{array}{ccc} & & (\mathbb{R}, 0) \\ & \overset{\alpha'}{\nearrow} & \downarrow p \\ (I, 0) & \xrightarrow{\alpha} & (S^1, 1) \end{array}$$

Let $\alpha : I \longrightarrow S^1$ be a path $\alpha(0) = 1$. By the local homeomorphism of p, there is a unique path $\alpha' : (I, 0) \longrightarrow (\mathbb{R}, 0), \alpha'(0) = 0$ such that $p \cdot \alpha' = \alpha$. If $H : I \times I \longrightarrow S^1$ is $H : \alpha \simeq \beta \quad \text{rel } (0,1)$, then there is $H' : I \times I \longrightarrow \mathbb{R}$ such that

$$H' : \alpha' \simeq \beta' \quad \text{rel } (0,1)$$

and

$$p \cdot H' = H.$$

In fact, since $I \times I$ is compact, H is uniformly continuous and p is a local homeomorphism, we can prove it.
This is a special case of the covering homotopy Theorem 8.2.3.

Theorem 8.3.1. $\pi_1(S^1) \simeq \mathbb{Z}$.

Proof. Consider a local diffeomorphism $p : \mathbb{R} \longrightarrow S^1$ defined by $p(t) = e^{2\pi it}$. For each $[\alpha] \in \pi_1(S^1, 1)$, there is a unique lifting $\alpha' : (I, 0) \longrightarrow (\mathbb{R}, 0)$ such that $\alpha'(0) = 0$, $p \circ \alpha' = \alpha$. Then

$$p(\alpha'(1)) = \alpha(1) = 1 \text{ and } \alpha'(1) \in p^{-1}(1) = \mathbb{Z}.$$

Define $\phi : \pi_1(S^1, 1) \longrightarrow \mathbb{Z} = p^{-1}(1)$ by

$$\phi([\alpha]) = \alpha'(1).$$

Let $[\alpha], [\beta] \in \pi_1(S^1, 1)$ and $\alpha'(1) = m, \beta'(1) = n$. Define

$$\beta'' : I \longrightarrow \mathbb{R}$$

by $\beta''(t) = m + \beta'(t)$. Then $\beta''(0) = m + \beta'(0) = m$, and for any $t \in \mathbb{R}$

$$\begin{aligned}(p \circ \beta'')(t) &= p(\beta''(t)) = p(m + \beta'(t)) \\ &= p(m) \cdot p(\beta'(t)) = 1 \cdot p(\beta'(t)) = (p \circ \beta')(t) \\ &= \beta(t).\end{aligned}$$

Thus $p \circ \beta'' = \beta$.

$$p \circ (\alpha' \circ \beta'') = (p \circ \alpha') \circ (p \circ \beta'')\alpha \circ \beta$$

and

$$\begin{aligned}\phi([\alpha][\beta]) &= \phi([\alpha \circ \beta]) = (\alpha' \circ \beta'')(1) = \beta''(1) \\ &= m + \beta'(1) = m + n = \phi([\alpha]) + \phi([\beta]).\end{aligned}$$

Therefore ϕ is a homomorphism.

For any $n \in \mathbb{Z}$, define $\alpha' : I \longrightarrow \mathbb{R}$ by $\alpha'(t) = nt$. Then $\alpha'(0) = 0$, $\alpha'(1) = n$. Define $\alpha : I \longrightarrow S^1$ by $\alpha = p \cdot \alpha'$. Then $[\alpha] \in \pi_1(S^1, 1)$ and $\phi([\alpha]) = \alpha'(1) = n$.
Thus ϕ is surjective.

Suppose $\phi([\alpha]) = 0$ and $[\alpha] = \pi_1(S^1, 1)$. Then there is a unique path $\alpha' : I \longrightarrow \mathbb{R}$ such that $\alpha'(0) = 0, p \circ \alpha' = \alpha$ and $\phi([\alpha]) = \alpha'(1) = 0$. $\alpha' \simeq 0 \quad \text{rel } (0, 1)$ in $\mathbb{R}$ by $H(s, t) = (1 - t)\alpha(s)$. So, $\alpha = p \circ \alpha' \simeq p \circ 0 = 1$ rel $(0, 1)$ in S^1. Then $[\alpha] = 1 \in \pi_1(S^1, 1)$.
Thus $\ker \phi = \{1\}$ and ϕ is one-to-one.
Hence $\phi : \pi_1(S^1, 1) \longrightarrow \mathbb{Z}$ is an isomorphism. □

Definition 8.3.2. For a subset $A \subset X$, A is a **retract** of X if there is a map $\gamma : X \longrightarrow A$ such that $\gamma(a) = a \ \forall a \in A$. In this case, γ is called a **retraction**.

Theorem 8.3.3. *The unit circle $S^1 \subset D^2$ is not a retract of D^2.*

Proof. If $\gamma : D^2 \longrightarrow S^1$ is a retraction, then $\gamma \circ i = I_{S^1}$ where $i : S^1 \longrightarrow D^2$ is the inclusion. By Theorem 8.2.3,

$$\gamma_* \circ i_* = (\gamma \circ i)_* = (I_{S^1})_* = I_{\pi_1(S^1)} : \mathbb{Z} \longrightarrow \mathbb{Z},$$

but $\pi_1(S^1) \xrightarrow{i_*} \pi_1(D^2) \xrightarrow{\gamma_*} \pi_1(S^1)$ is a contradiction ($\pi_1(S^1) = \mathbb{Z}$ and $\pi_1(D^2) = \{1\}$).
Thus there is no retraction $\gamma : D^2 \longrightarrow S^1$. □

Theorem 8.3.4. $\pi_1(X \times Y, (x_o, y_0)) \simeq \pi_1(X, x_o)) \times \pi_1(Y, y_o))$.

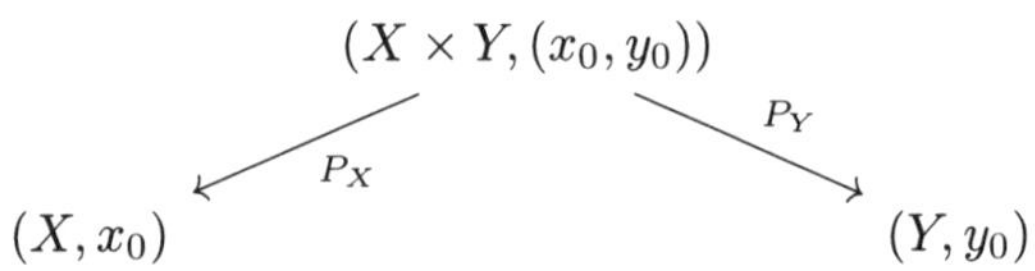

Proof. Let $(X \times Y, (x_0, y_0))$. Define

$$\phi : \pi_1(X \times Y, (x_o, y_0)) \longrightarrow \pi_1(X, x_0) \times \pi_1(Y, y_0)$$

by $\phi([\alpha]) = (P_{X_*}[\alpha], P_{Y_*}[\alpha]) = ([P_X \circ \alpha], [P_Y \circ \alpha])$. Then ϕ is a homomorphism. For every $([\alpha_1], [\alpha_2]) \in \pi_1(X, x_o)) \times \pi_1(Y, y_o))$ define $\alpha : I \longrightarrow X \times Y$ by $\alpha(t) = (\alpha_1(t), \alpha_2(t))$. Then $\alpha(0) = (\alpha_1(0), \alpha_2(0))$ $= (x_0, y_0) = \alpha(1)$. Define

$$\Psi : \pi_1(X, x_o)) \times \pi_1(Y, y_0)) \longrightarrow \pi_1(X \times Y, (x_o, y_0))$$

by $\Psi([\alpha_1, \alpha_2]) = [\alpha]$. Then Ψ is the inverse of ϕ.
Thus $\phi : \pi_1(X \times Y, (x_0, y_0)) \longrightarrow \pi_1(X, x_0) \times \pi_1(Y, y_0)$ is an isomorphism. □

Example 8.3.5. (1) $\pi_1(S^1 \times S^1) \simeq \pi_1(S^1) \times \pi_1(S^1) \simeq \mathbb{Z} \times \mathbb{Z}$.

(2) $\pi_1(S^1 \times D^2) \simeq \pi_1(S^1) \times \pi_1(D^2) \simeq \mathbb{Z} \times \{1\} \simeq \mathbb{Z}.$

$S^1 \times S^1 =$ 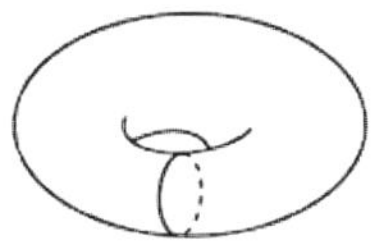 $S^1 \times D^2 =$ 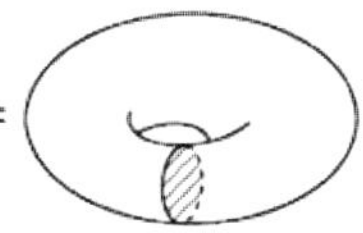

Thus $S^1 \times S^1 \not\simeq S^1 \times D^2 \simeq S^1$.

8.4 Seifert-Van Kampen Theorem

Theorem 8.4.1 (Seifert-Van Kampen Theorem). *Let $X = X_1 \cup X_2$, $X_1, X_2, X_1 \cap X_2$ be path connected, and $x_0 \in X_1 \cap X_2$. Then there is an isomorphism*

$$\pi_1(X, x_0) \simeq \frac{\pi_1(X, x_0) * \pi_1(X_2, x_0)}{H},$$

where $*$ is the free product,

$$H = N\{i_{1_*}\alpha \cdot (i_{2_*}\alpha)^{-1} \mid \alpha \in \pi_1(X_1 \cap X_2, x_0)\}$$

and

$$\begin{array}{ccc} X_1 \cap X_2 & \xrightarrow{i_1} & X_1 \\ \downarrow{\scriptstyle i_2} & & \downarrow{\scriptstyle j_2} \\ X_2 & \xrightarrow{j_1} & X, \end{array}$$

are inclusions, and induced maps

$$\begin{array}{ccc} \pi_1(X_1 \cap X_2, x_0) & \xrightarrow{i_{1*}} & \pi_1(X_1, x_0) \\ \downarrow{\scriptstyle i_{2*}} & & \downarrow{\scriptstyle j_{1*}} \\ \pi_1(X_2, x_0) & \xrightarrow{j_{2*}} & \pi(X, x_0) \end{array}$$

are commutative.

Remark 8.4.2. (1) $\frac{\pi_1(X,x_0)*\pi_1(X_2,x_0)}{H}$ is called the **amalgamated free product** of $\pi_1(X_1, x_0)$ and $\pi_1(X_2, x_0)$ with respect to H.

(2) $[\alpha] = [\alpha_1 \cdot \alpha_2 \cdot \alpha_3] \in \pi_1(X, x_0)$

$[\alpha_1 \cdot {\beta_1}^{-1}] \in \pi_1(X_1, x_0)$

$[{\beta_2}^{-1} \cdot \alpha_3] \in \pi_1(X_2, x_0)$

$[\beta_1 \cdot \alpha_2 \cdot \beta_2] \in \pi_1(X_1 \cap X_2, x_0)$

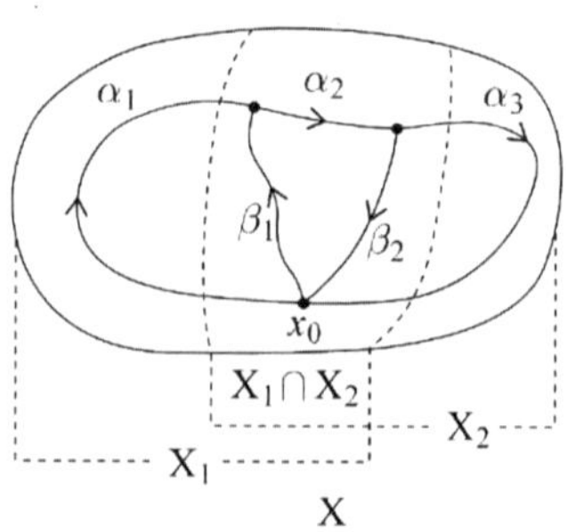

(3) In Theorem 8.4.1, given diagram

$$\begin{array}{ccc} \pi_1(X_1 \cap X_2, x_0) & \xrightarrow{i_{1*}} & \pi_1(X_1, x_0) \\ \downarrow{\scriptstyle i_{2*}} & & \\ \pi_1(X_2, x_0) & & \end{array}$$

the group $\pi_1(X, x_0) \simeq \frac{\pi_1(X,x_0) * \pi_1(X_2,x_0)}{H}$ is the pushout of the given diagram, which is uniquely determined by the universality, that is,

$$\begin{array}{ccccc} \pi_1(X_1 \wedge X_2, x_0) & \xrightarrow{i_{1*}} & \pi_1(X_1, x_0) & & \\ \downarrow{\scriptstyle i_{2*}} & & \downarrow{\scriptstyle j_{1*}} & \searrow{\scriptstyle g_1} & \\ \pi_1(X_2, x_0) & \xrightarrow{j_{2*}} & \pi_1(X, x_0) & \overset{h_1}{\dashrightarrow} & G \\ & \searrow{\scriptstyle g_2} & & & \end{array}$$

(i) $j_{1*} \cdot i_{1*} = j_{2*} \cdot i_{2*}$

(ii) for any homomorphisms g_1, g_2 and any group G such that $g_1 \circ i_{1*} = g_2 \circ i_{2*}$, there is a unique homomorphism

$$h : \pi_1(X, x_0) \longrightarrow G$$

such that $g_1 = h \circ j_{1*}$, $g_2 = h \circ j_{2*}$. □

Corollary 8.4.3. (1) *If $X_1 \cap X_2$ is simply connected, then*

$$\pi_1(X, x_0) \simeq \pi_1(X_1, x_0) * \pi_1(X_2, x_0).$$

(2) *If $\pi_1(X_2, x_0) = \{1\}$ i.e., X_2: simply connected, then*

$$\pi_1(X, x_0) \simeq \frac{\pi_1(X_1, x_0)}{H}$$

where H is the normal subgroup generated by $im(i_{1_*})$ in $\pi_1(X, x_0)$.

Example 8.4.4. Let $X = S^1 \vee S^1$ be the figure 8. Then $\pi_1(X) \simeq \mathbb{Z} * \mathbb{Z}$.

Solution : Let $X = X_1 \vee X_2$, $X_1 = X - \{x_1\}$, $X_2 = X - \{x_2\}$

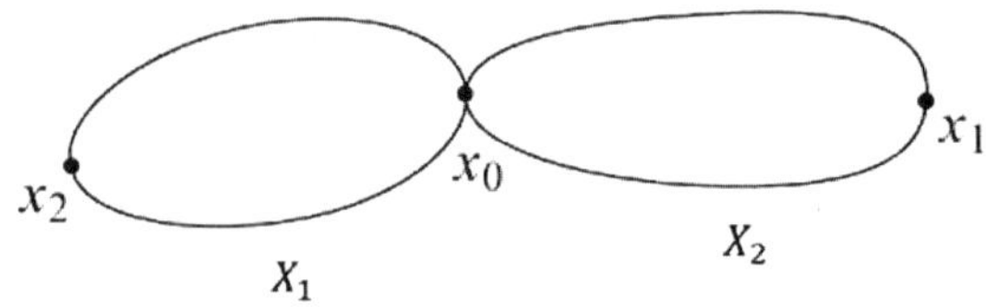

and choose $x_0 \in X_1 \cap X_2$. Then

$$\pi_1(X_1 \cap X_2, x_0) = \pi_1(x_0) = \{1\},$$

$$\pi_1(X_1, x_0) = \pi_1(S^1, x_0) = \mathbb{Z}$$
$$\pi_1(X_2, x_0) = \pi_1(S^1, x_0) = \mathbb{Z}$$

Thus by Corollary 8.4.3,

$$\pi_1(X, x_0) \simeq \pi_1(X_1, x_0) * \pi_1(X_2, x_0)$$
$$= \mathbb{Z} * \mathbb{Z}$$

□

8.5 Examples of Fundamental Group

Recall the following examples:

(1) The unit sphere :

$$S^2 = \{(x_1, x_2, x_3) \in \mathbb{R}^3 \mid {x_1}^2 + {x_2}^2 + {x_3}^2 = 1\} =$$

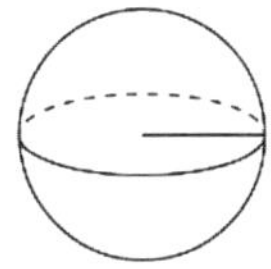

(2) The torus :

$$T^2 = S^1 \times S^1 = \mathbb{R}^2/\mathbb{Z}^2 =$$

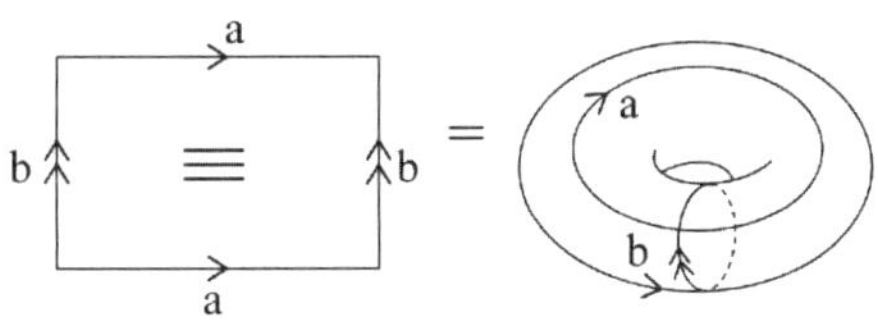

(3) The real projective plane :

$$\mathbb{RP}^2 = S^2/\sim\ =$$

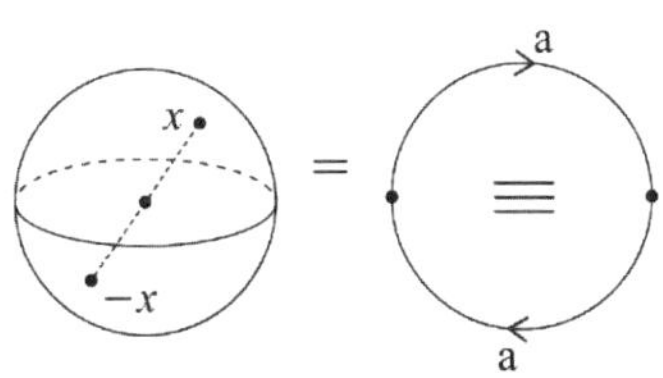

where $x \sim y \Leftrightarrow x = y$ or $x = -y$.

Definition 8.5.1. Let S_1 and S_2 be two surfaces and $D_1 \subset S_1$, $D_2 \subset S_2$ small discs. Let $h : \partial D_1 \longrightarrow \partial D_2$ be a homeomorphism between two circles.

$$S_1 \sharp S_2 := (S_1 - \mathring{D}_1) \amalg (S_2 - \mathring{D}_2) /_{\sim}$$

where $x \sim h(x)$ for each $x \in \partial D_1$.
Then $S_1 \sharp S_2$ is called the **connected sum** of S_1 and S_2.

Example 8.5.2. (1) $S^2 \sharp S \simeq S$ for any surface S.

(2) 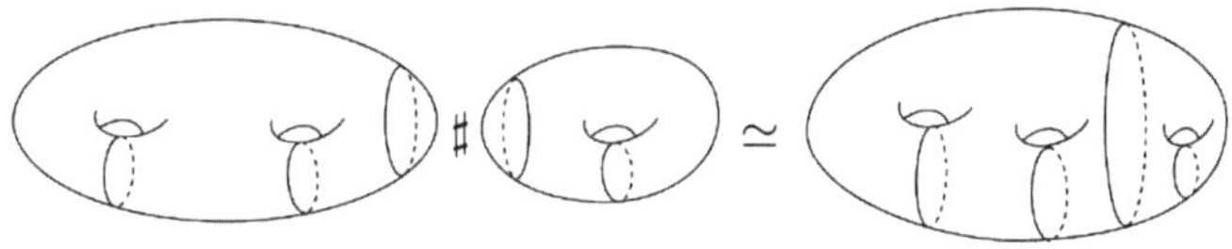

(3) $T^2 \sharp \mathbb{RP}^2 \simeq \mathbb{RP}^2 \sharp \mathbb{RP}^2 \sharp \mathbb{RP}^2$.

Theorem 8.5.3. *If X is a compact, connected surface, then X is either S^2 or $\overset{n}{\sharp}\Gamma^2$ or $\overset{m}{\sharp}\mathbb{RP}^2$ for some n, m, where $\overset{n}{\sharp}$ means the connected sum of n copies. The surfaces S^2 or $\overset{n}{\sharp}T^2$ are orientable and $\overset{m}{\sharp}\mathbb{RP}^2$ are not.*

Recall the followings:

(1) $\pi_1(S^1) = \mathbb{Z} = \langle a_1 \rangle$, 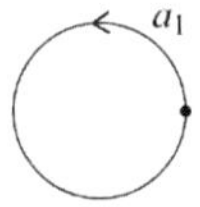

(2) $\pi_1(\vee_n S^1) = \mathbb{Z} * \cdots * \mathbb{Z} = \langle a_1 \rangle * \cdots * \langle a_n \rangle$

where $\vee_n S^1 =$ 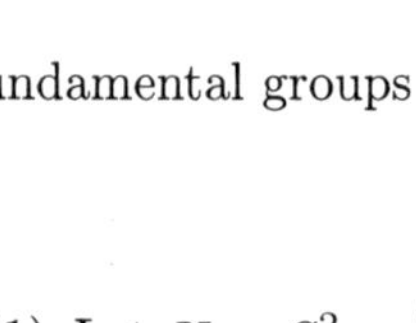 is an n-bouquet.

Compute the fundamental groups of surfaces.

Example 8.5.4. (1) Let $X = S^2 =$ 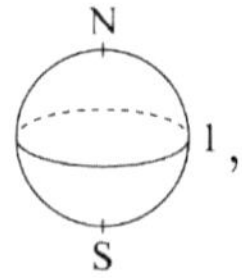

, $X_1 = X - \{N\} \simeq \text{pt}$, $X_2 = X - \{S\} \simeq \text{pt}$.

By Theorem 8.4.1,

$$\pi_1(S^2) \simeq \frac{\pi_1(X_1) * \pi_1(X_2)}{H} = \{1\}.$$

(2) Let $X = T^2 =$ 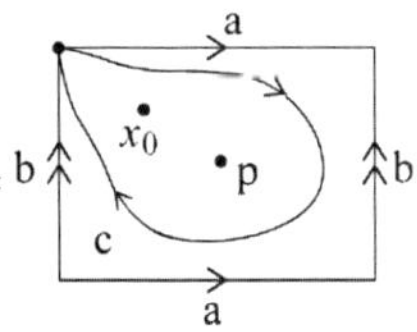

be a torus, and $X_1 = X - \{p\}$, $X_2 = X - \{a, b\}$. Then

$X_1 \simeq$ 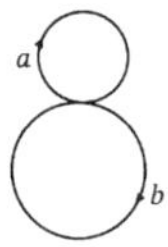

, $\quad X_2 \simeq \text{pt}, \quad X_1 \cap X_2 \simeq c.$

$$\begin{array}{ccc} X_1 \cap X_2 & \xrightarrow{i_1} & X_1 \\ \big\downarrow{\scriptstyle i_2} & & \big\downarrow{\scriptstyle j_1} \\ X_2 & \xrightarrow{j_2} & X, \end{array} \qquad \begin{array}{ccc} \pi_1(X_1 \cap X_2) & \xrightarrow{i_{1*}} & \pi_1(X_1) \\ \big\downarrow{\scriptstyle i_{2*}} & & \big\downarrow{\scriptstyle j_{1*}} \\ \pi_1(X_2) & \xrightarrow{j_{2*}} & \pi_1(X) \end{array}$$

then $\pi_1(X_1) = \langle [a] \rangle * \langle [b] \rangle$, $\pi_1(X_2) = \{1\}$ $\pi_1(X_1 \cap X_2) = \langle [c] \rangle$, and $i_{1*}[c] = [aba^{-1}b^{-1}] \in \pi_1(X_1)$. Thus

$$\begin{aligned} \pi_1(X) &= \frac{\pi_1(X_1) * \pi_1(X_2)}{N(\operatorname{Im} i_{1*})} = \frac{\pi_1(X_1)}{N(\operatorname{Im} i_{1*})} \\ &= \frac{\langle [a] \rangle * \langle [b] \rangle}{[aba^{-1}b^{-1}] \rangle} = \langle [a] \rangle \times \langle [b] \rangle = \mathbb{Z} \times \mathbb{Z}. \end{aligned}$$

(3) Let $X = \mathbb{RP}^2 =$ 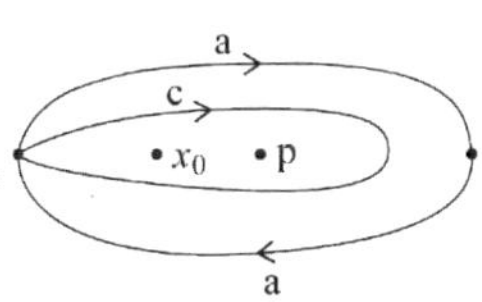

: Real projective plane

$$X_2 = X - \{a\}, \quad X_1 = X - \{p\}$$

Then $X_1 \simeq a$, $X_2 \simeq \text{pt}$, and $X_1 \cap X_2 \simeq c$. Consider

$$\begin{array}{ccc} \pi_1(X_1 \cap X_2) & \xrightarrow{i_{1*}} & \pi_1(X_1) \\ \big\downarrow{\scriptstyle i_{2*}} & & \big\downarrow{\scriptstyle j_{1*}} \\ \pi_1(X_2) & \xrightarrow{j_{2*}} & \pi_1(X) \end{array}$$

Then $\pi_1(X_1 \wedge X_2) = \langle [c] \rangle = \mathbb{Z}$, $\pi_1(X_1) = \langle [a] \rangle = \mathbb{Z}$, $\pi_1(X_2) = \{1\}$, and $i_{1*}[c] = [a^2]$. Thus

$$\pi_1(X) = \frac{\pi_1(X_1)}{N(\operatorname{Im} i_{1*})} = \frac{\langle [a] \rangle}{\langle [a]^2 \rangle} = \mathbb{Z}_2.$$

(4) Let $X = \overset{n}{\sharp} T^2 =$ 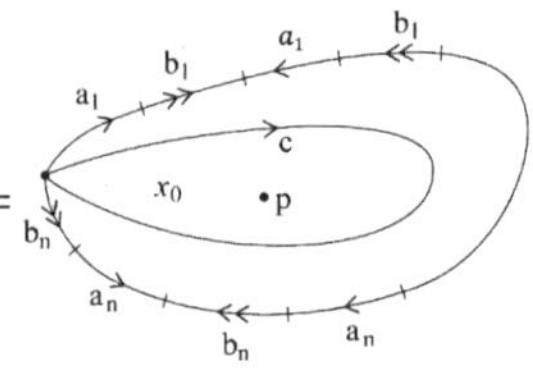

: the connected sum of n tori

: the surface with genus n.

$$X_1 = X - \{p\}, \quad X_2 = X - \{a_1, b_1, \cdots a_n, b_n\} \simeq \text{pt}$$
$$\simeq (\vee_n a_i) \vee (\vee_n b_i) \simeq 2n\text{-bouquet}$$

$X_1 \cap X_2 \simeq c$. Then

$$\pi_1(X_1) = \langle [a_1] \rangle * \langle [b_1] \rangle * \cdots * \langle [a_n] \rangle * \langle [b_n] \rangle$$
$$\pi_1(X_2) = \{1\}, \pi_1(X_1 \cap X_2) = \langle [c] \rangle$$
$$i_{1*}([c]) = [a_1 b_1 a_1^{-1} b_1^{-1} \cdots a_n b_n a_n^{-1} b_n^{-1}].$$

Thus by Theorem 8.3.1,

$$\pi_1(X) = \frac{\langle [a_1] \rangle * \langle [b_1] \rangle * \cdots * \langle [a_n] \rangle * \langle [b_n] \rangle}{N([a_1 b_1 a_1^{-1} b_1^{-1} \cdots a_n b_n a_n^{-1} b_n^{-1}])}$$

(5) Let $X = \overset{n}{\sharp} \mathbb{RP}^2 =$ 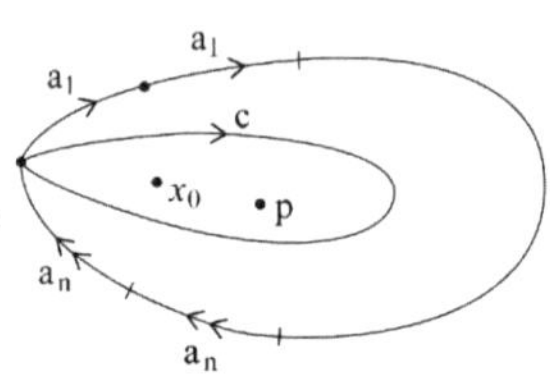

$$X_1 = X - \{p\} \simeq \vee_n a_i,$$
$$X_2 = X - \{a_1, b_1, \cdots a_n\} \simeq \text{pt}, \quad \text{and}$$
$$X_1 \cap X_2 \cong c$$

Thus $\pi_1(X_1) = \langle [a_1] \rangle * \cdots * \langle [a_n] \rangle$, $\pi_1(X_2) = \{1\}$,

$$\pi_1(X_1 \cap X_2) = \langle [c] \rangle, \quad i_{1*}([c]) = [{a_1}^2 \cdots {a_n}^2].$$

By Theorem 8.3.1, we have

$$\pi_1(X) = \frac{\langle [a_1] \rangle * \cdots * \langle [a_n] \rangle}{N([{a_1}^2 \cdots {a_n}^2])}.$$

Chapter 9

Covering Spaces

9.1 Covering Spaces

Definition 9.1.1. A space X is **locally pathwise connected** if for each point $x \in X$ and any nbd U of x, there is a nbd V of x such that $x \in V \subset U$ and for each points $x_0, x_1 \in V$, there is a path $\gamma : I \longrightarrow V$ with $\gamma(0) = x_0$, $\gamma(1) = x_1$.

Remark 9.1.2. X is pointwise connected $\nRightarrow$ X is locally pathwise connected.

For example, $X = \{(0, y) \mid y \in [0, 1]\} \cup \{(x, \sin\frac{1}{x}) \mid x \in (0, 1]\}$
$\cup\{(x, 0) \mid x \in [0, 1]\} := X_1 \cup X_2 \cup X_3$, respectively.

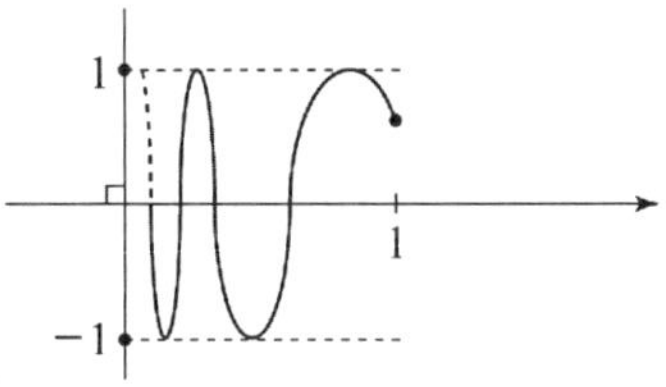

(1) $X_1 \cup X_2$ is connected, not (locally) pathwise connected.

(2) X is connected and pathwise connected, but not locally pathwise connected.

(3) $[0, 1] \cup [2, 3]$ is locally pathwise connected, but not connected.

Definition 9.1.3. Let $X, \widetilde{X}$ be pathwise connected and locally pathwise connedted. A map $p : \widetilde{X} \longrightarrow X$ is a **covering space** of X if

(1) p is onto

(2) For each point $x \in X$, there is a nbd of x, U such that

$$p^{-1}(U) = \coprod_\alpha S_\alpha \text{ is disjoint union of open sets, and for each } \alpha \\ \text{restriction } p : S_\alpha \longrightarrow U \text{ is a homeomorphism}$$

Example 9.1.4. (1) Let $X = \widetilde{X} = S^1$. Then the map

$$p : \widetilde{X} \longrightarrow X, \quad p(z) = z^n$$

is an n-fold covering space.

$$p^{-1}((1, 0)) = \{e^{\frac{2\pi ik}{n}} \mid k = 0, 1, \cdots n - 1\}.$$

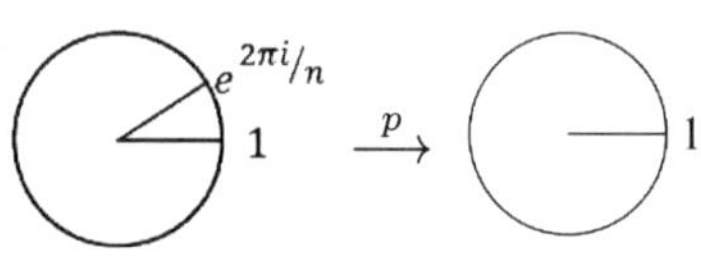

(2) Let $X = S^1$, $\widetilde{X} = \mathbb{R}$. Then the map

$$p : \widetilde{X} \longrightarrow X, \quad p(t) = e^{2\pi it}$$

is a covering space.

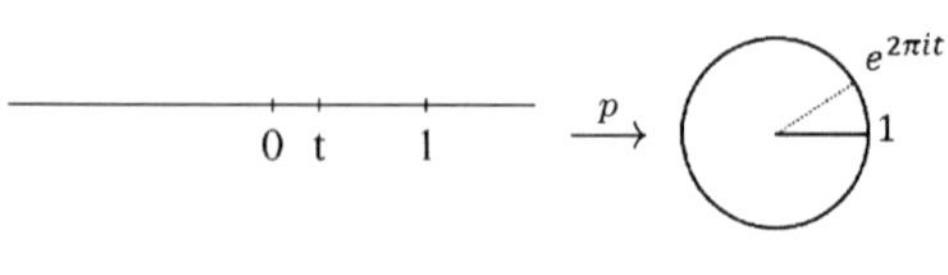

$$p^{-1}((1, 0)) = \{\cdots, -1, 0, 1, \cdots\} = \mathbb{Z}.$$

(3) $X = \mathbb{RP}^2$, $\widetilde{X} = S^2$. The map

$$p : S^2 \longrightarrow \mathbb{RP}^2, \quad x \mapsto p(x) = [x] = \{x, -x\}$$

is a double covering space.

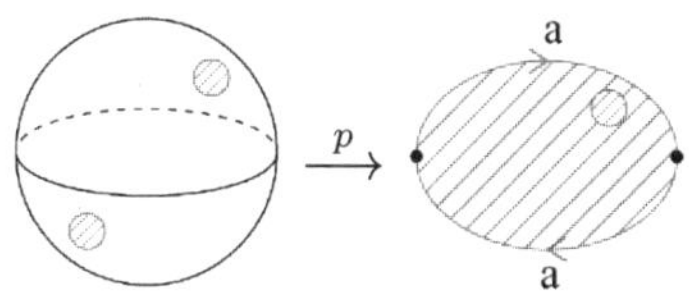

$$p^{-1}([x]) = \{x, -x\}.$$

(4) Let $X = S^1 \times S^1$, $\widetilde{X} = \mathbb{R}^2$. The map

$$p : \mathbb{R}^2 \longrightarrow S^1 \times S^1 \quad p(s,t) = (e^{2\pi is}, e^{2\pi it})$$

is a covering space.

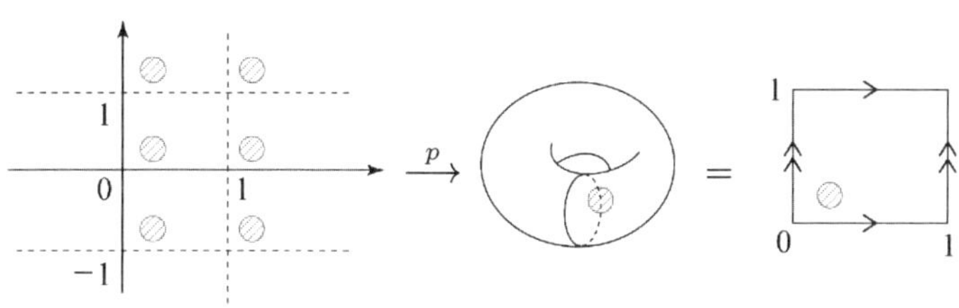

$$p^{-1}((1,1)) = \mathbb{Z} \times \mathbb{Z},$$

where $1 = (1,0)$.

Definition 9.1.5. Let $p : (\widetilde{X}, \widetilde{x_0}) \longrightarrow (X, x_0)$ be a covering space, $p(\widetilde{x_0}) = x_0$. For a map $f : Y \longrightarrow X$ the map $f' : Y \longrightarrow \widetilde{X}$ such that $p \circ f' = f$ is called a lifting of f.

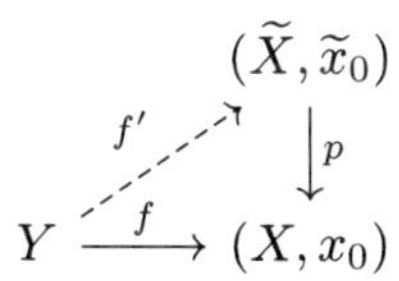

In Example 9.1.4 (2), each path $\alpha : I \longrightarrow S^1$ has a lifting $\alpha' : I \longrightarrow \mathbb{R}$, i.e., $\alpha' \circ p = \alpha$. If $\alpha'(0) = n$, then α' is unique.

9.2 Covering Homotopy Theorem

Theorem 9.2.1. *Let* $p : (\widetilde{X}, \widetilde{x_0}) \longrightarrow (X, x_0)$ *be a covering space,* $f : (Y, y_0) \longrightarrow (X, x_0)$ *a map, and* Y *connected. If* $f' : (Y, y_0) \longrightarrow (\widetilde{X}, \widetilde{x_0})$ *is a lifting of* f*, then* f' *is unique.*
(i.e., If $f' \circ p = f$ *and* $f'(y_0) = \widetilde{x_0}$*, then* f' *is unique.)*

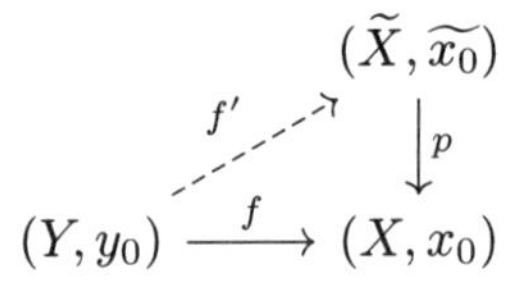

Proof. Suppose that $f'' : (Y, y_0) \longrightarrow (\widetilde{X}, \widetilde{x_0})$, $p \circ f'' = f$, is another lifting of f. Let $Y_1 = \{y \in Y \mid f'(y) = f''\}$, $Y_2 = \{y \in Y \mid f'(y) \neq f''\}$. Then

$$Y = Y_1 \cup Y_2, \quad Y_1 \cap Y_2 = \phi, \quad y_0 \in Y_1.$$

Show $Y_1 = Y$ (i.e., Y_1, Y_2 are open in Y).

If $y_1 \in Y_1$, then $f(y_1) \in X$ has a nbd U such that $p^{-1}(U) = \coprod_\alpha S_\alpha$ is a disjoint union of open sets S_α and $p : S_\alpha \longrightarrow U$ us a homeomorphism for each α. Since $f'(y_1) = f''(y_1)$ and $p(f'(y_1)) = f(y_1) \in U$, there is a unique α such that $f'(y_1) = f''(y_1) \in S_\alpha$. Then $f'^{-1}(S_\alpha) \cap f''^{-1}(S_\alpha)$ is a nbd of y_1. Since $p : S_\alpha \longrightarrow U$ is a homeomorphism $f^{-1'}(S_\alpha) \cap f^{-1''}(S_\alpha) \subset Y_1$ Thus Y_1 is open.

If $y_2 \in Y_2$, then there is an open nbd V of $f(y_2)$ in Y such that $f^{-1}(V) = \coprod_\beta S_\beta$ is a disjoint union of open S_β and $p : S_\beta \longrightarrow V$ is a homeomorphism for each β. There are $\beta_1 \neq \beta_2$ such that $f^{-1}(y_2) \in S_{\beta_1}$ and $f^{-1''}(y_2) \in S_{\beta_2}$. So $f'^{-1}(S_{\beta_2}) \cap f''^{-1}(S_{\beta_2})$ is a nbd of y_2 and contained in Y_2.

Thus Y_2 is also open.

Since $Y = Y_1 \amalg Y_2$ and Y_1 is closed and open nonempty and Y is connected, $Y_1 = Y$. □

Theorem 9.2.2. *Let $p : (\widetilde{X}, \widetilde{x_0}) \longrightarrow (X, x_0)$ be a covering space. Then a path $\alpha : (I, 0) \longrightarrow (X, x_0)$ has a unique lifting $\alpha' : (I, 0) \longrightarrow (\widetilde{X}, \widetilde{x_0})$ of α.*

Proof. Since I is a compact metric space and $p : \widetilde{X} \longrightarrow X$ is a covering space, there is a partition of I, $0 = t_0 < t_1 < \cdots < t_n = 1$, such that $\alpha(t_i, t_{i+1}) \subset$ a nbd of U_i of $\alpha(t_i)$ and $p^{-1}(U_i) = \coprod_j S_{ij}$ is a disjoint union of open S_{ij} and $p : S_{ij} \longrightarrow U_i$ is a homeomorphism.

Define

$$\Big[\alpha_1' : ([0, t_1], 0) \longrightarrow (\widetilde{X}, \widetilde{x_0})\Big] := \Big[([0, t_1], 0) \xrightarrow{\alpha} (U_0, x_0) \xrightarrow{p^{-1}} (S_0, \widetilde{x_0})\Big]$$

and

$$\begin{aligned}&\Big[\alpha_2' : \big([t_1, t_2], t_1\big) \longrightarrow (\widetilde{X}, \alpha_1'(t_1))\Big]\\ &\quad := \Big[\big([t_1, t_2], t_1\big) \xrightarrow{\alpha} (U_1, \alpha(t_1)) \xrightarrow{p^{-1}} (S_1, \alpha_2'(t_1))\Big]\end{aligned}$$

and finally we get

$$\alpha_n' = \Big[\big([t_{n-1}, t_n], t_{n-1}\big) \xrightarrow{\alpha} (U_{n-1}, \alpha(t_{n-1})) \xrightarrow{p^{-1}} (S_{n-1}, \alpha_{n-1}'(t_{n-1}))\Big].$$

By Theorem 9.2.1, the path $\alpha' = \alpha_1' \cdot \alpha_2' \cdots \alpha_n' : ([0,1], 0) \longrightarrow (\widetilde{X}, \widetilde{x_0})$ is the unique lifting of α. □

Theorem 9.2.3 (Covering Homotopy Theorem). *Let*

$$p : (\widetilde{X}, \widetilde{x_0}) \longrightarrow (X, x_0)$$

be a covering space. If $f : (Y, y_0) \longrightarrow (X, x_0)$ has a lifting

$$f' : (Y, y_0) \longrightarrow (\widetilde{X}, \widetilde{x_0}),$$

then a homotopy

$$F : Y \times I \longrightarrow X, \quad F(y, 0) = f(y) \;\; \forall y \in Y$$

has a lifting homotopy $F' : Y \times I \longrightarrow \widetilde{X}$,

$$F'(y,0) = f'(y) \quad \forall y \in Y$$

and

$$p \circ F' = F.$$

Proof. Essentially, the proof is the same one as Theorem 9.2.2.

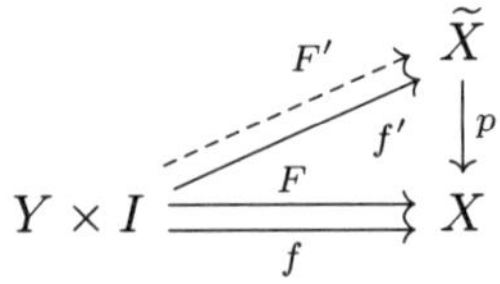

For each point $y \in Y$, there is a nbd N_y of y in X and a partition $0 = t_0 < t_1 < \cdots < t_n = 1$ such that $F(N_y \times [t_i, t_{i+1}])$ is contained in a covering nbd of $F(y, t_i)$ in X.

As in the proof of Theorem 9.2.2 the homotopy $F : N_y \times I \longrightarrow X$ has a lifting homotopy $F' : N_y \times I \longrightarrow \widetilde{X}$ and $\widetilde{F}(y', 0) = f'(y')\ \forall y' \in N_y$.

If $y \in N_{y_1} \cap N_{y_2}$, then $F\mid_{y \times I}$ has two lifting $F'\mid_{y_1 \times I}$, $F'\mid_{y_2 \times I}$ and they are same at $(y, 0)$. Since $y \times I$ is connected, by Theorem 9.2.1 the two lifting are the same on $y \times I$.

Thus two lifting are the same on $(N_{y_1} \cap N_{y_2}) \times I$.

Hence the lifting homotopy $F' : Y \times I \longrightarrow \widetilde{X}$ of F is well-defined such that $F'(y,0) = f'(0)\ \forall y \in Y$. □

Theorem 9.2.4. *Let* $p : (\widetilde{X}, \widetilde{x_0}) \longrightarrow (X, x_0)$ *be a covering space. If* $\alpha \simeq \beta$ rel $(0,1)$: $(I, 0) \longrightarrow (\widetilde{X}, \widetilde{x_0})$, *then*

$$\alpha' \simeq \beta' \text{rel } (0,1) : (I,0) \longrightarrow (\widetilde{X}, \widetilde{x_0}).$$

In particular, $\alpha'(1) = \beta'(1)$.

Proof. By Theorem 9.2.2, there is lifting α' of α. By Theorem 9.2.3, there is covering homotopy, and by Theorem 9.2.1, $\alpha' \simeq \beta'$ rel $(0,1)$. Thus $\alpha'(1) = \beta'(1)$. □

Corollary 9.2.5. *Let $p : (\widetilde{X}, \widetilde{x_0}) \longrightarrow (X, x_0)$ be a covering space. Then $p_* : \pi_1(\widetilde{X}, \widetilde{x_0}) \longrightarrow \pi_1(X, x_0)$ is one-to-one.*

Proof. If $p_*([\alpha']) = p_*([\beta'])$, then $p \circ \alpha' \simeq p \circ \beta'$ rel $(0,1)$. By Theorem 9.2.4, $\alpha' \simeq \beta'$ rel $(0,1)$ and $[\alpha'] = [\beta']$ □

Corollary 9.2.6. *Let $p : (\widetilde{X}, \widetilde{x_0}) \longrightarrow (X, x_0)$ be a covering space and $(\widetilde{X}, \widetilde{x_0})$ path connected. Then the map $c : \pi_1(X, x_0) \longrightarrow p^{-1}(x_0)$, $c([\alpha]) = \alpha'(1)$ is well-defined and the induced map*

$$\bar{c} : \pi_1(X, x_0) \Big/ p_*\pi_1(\widetilde{X}, \widetilde{x_0}) \longrightarrow p^{-1}(x_0),$$

is bijective. Here α' is a lifting of α and $\alpha'(0) = \widetilde{x_0}$.

Proof. (i) Since X is path connected, c is onto: For any $\widetilde{x} \in p^{-1}(x_0)$ choose a path $\alpha' : [0,1] \longrightarrow \widetilde{X}$ such that $\alpha'(0) = \widetilde{x_0}$, $\alpha'(1) = \widetilde{x}$. Let $\alpha = p \circ \alpha' : I \longrightarrow X$. Then $\alpha(0) = \alpha(1) = x_0$ and $[\alpha] \in \pi_1(X, x_0)$. By definition, $c([\alpha]) = \alpha'(1) = \widetilde{x}$.

(ii) The map

$$\bar{c} : \pi_1(X, x_0)/p_*\pi_1(\widetilde{X}, \widetilde{x_0}) \longrightarrow p^{-1}(x_0)$$

is well-defined and bijective. Since $c^{-1}(\widetilde{x_0}) = p_*\pi_1(\widetilde{X}, \widetilde{x_0})$, $\bar{c}$ is well-defined and onto. If $\bar{c}([\alpha]) = \bar{c}([\beta])$, then $\alpha'(1) = \beta'(1)$, and $[\alpha' \cdot \beta'^{-1}] \in \pi_1(\widetilde{X}, \widetilde{x_0})$. So

$$\begin{aligned} p_*[\alpha' \cdot \beta'^{-1}] &= [p \circ \alpha' \cdot p \circ \beta'^{-1}] \\ &= [\alpha \cdot \beta^{-1}] = [\alpha] \cdot [\beta]^{-1} \in p_*\pi_1(\widetilde{X}, \widetilde{x_0}), \end{aligned}$$

and $[\alpha] \cdot p_*\pi_1(\widetilde{X}, \widetilde{x_0}) = [\beta] \cdot p_*\pi_1(\widetilde{X}, \widetilde{x_0})$.
Thus $\bar{c}$ is one-to-one.

□

Remark 9.2.7. (1) If $p : \mathbb{R} \longrightarrow S^1$, $p(t) = e^{2\pi i t}$, then $\pi_1(\mathbb{R}^1) = \{1\}$.

$$p^{-1}(1) = \mathbb{Z} \simeq \pi_1(S^1).$$

(2) If $p : S^2 \longrightarrow \mathbb{RP}^2$, $p(x) = [x]$, then $\pi_1(S^2) = \{1\}$.

$$p^{-1}(\mathrm{pt}) = \pi_1(\mathbb{RP}^2) \simeq \mathbb{Z}_2.$$

(3) If $p : \mathbb{R}^2 \longrightarrow S^1 \times S^1$, $p(s,t) = (e^{2\pi i s}, e^{2\pi i t})$, then $p^{-1}(1,1) = \mathbb{Z} \times \mathbb{Z} \simeq \pi_1(S^1 \times S^1)$, since $\pi_1(\mathbb{R}^2) = \{1\}$.

(4) If $p : S^1 \longrightarrow S^1$, $p(z) = z^n$, then $\pi_1(1) = \{e^{\frac{2\pi i k}{n}} \mid k = 0, \cdots, n-1\}$ $\simeq \mathbb{Z}_n \simeq \frac{\pi_1(S^1)}{p_*\pi_1(S^1)} = \mathbb{Z}/n\mathbb{Z}$.

9.3 Covering Spaces and Fundamental Groups

Definition 9.3.1. (1) A covering space $p : \widetilde{X} \longrightarrow X$ is a **universal covering spaces** if $\pi_1(\widetilde{X}) = \{1\}$ i.e., $\widetilde{X}$ is simply connected.

(2) X is **locally simply connected** if for each point $x \in X$, there is a nbd V of x such that any loop $\alpha : I \longrightarrow V$, $\alpha(0) = \alpha(1) = x$ is homotopic to the constant loop x in X.

Theorem 9.3.2. *Let X be path connected, locally path connected and locally simply connected. Then*

(1) *for a subgroup $H \subset \pi_1(X)$, there is a covering space $p : (\widetilde{X}, \widetilde{x_0}) \longrightarrow (X, x_0)$ such that $p_*\pi_1(\widetilde{X}, \widetilde{x_0}) = H$.*

(2) *If $H = \{1\}$, then $p : \widetilde{X} \longrightarrow X$ is a universal covering space.*

Proof. Fix a point $x_0 \in X$. Let

$$\Omega(X, x_0) := \{\alpha : I \longrightarrow X \text{ is a path} \mid \alpha(0) = x_0\}.$$

Define a relation $\underline{H}$ on $\Omega(X, x_0)$ by ${}_\alpha\underline{H}_\beta \Leftrightarrow \alpha(1) = \beta(1)$ and $[\alpha{\cdot}\beta^{-1}] \in H$. Then $\underline{H}$ is an equivalence relation on $\Omega(X, x_0)$. Define

$$\widetilde{X} := \Omega(X, x_0)/\underline{H}$$

and $[\alpha] := H\alpha$. For each $[\alpha] \in \widetilde{X}$ and U a nbd of $\alpha(1)$, define

$$([\alpha], U) := \{[\alpha, \beta] \mid \beta : I \longrightarrow U \text{ is a path}, \quad \beta(0) = \alpha(1)\}.$$

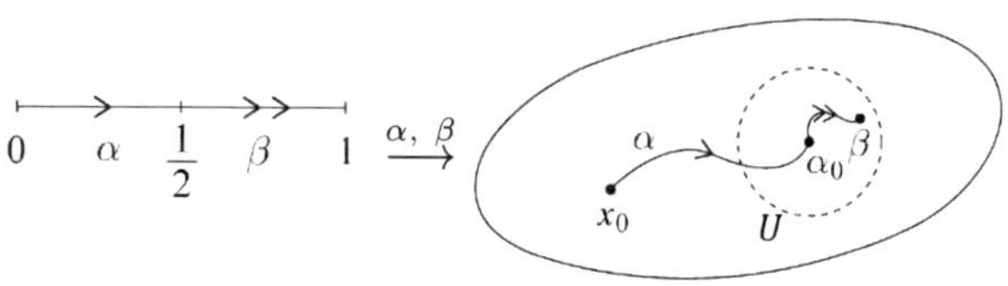

Then

$$\mathfrak{B}_{\widetilde{X}} = \{\phi, ([\alpha], U) \mid [\alpha] \in \widetilde{X}, \; U \text{ is a nbd of } \alpha(1)\}$$

is a basis for a topology of $\widetilde{X}$. Also $\widetilde{X}$ is path connected. Let $\widetilde{x_0} = hx_0 = [x_0] \in \widetilde{X}$. Then the map

$$p : \widetilde{X} \longrightarrow X \text{ given by } p([\alpha]) = \alpha(1)$$

is continuous, surjective and $p(x_0) = x_0$.
Thus $p : (\widetilde{X}, \widetilde{x_0}) \longrightarrow (X, x_0)$ is a covering space and $p_*\pi_1(\widetilde{X}, \widetilde{x_0}) = H$.

□

Theorem 9.3.3. *Let X be locally simply connected,*

$$p : (\widetilde{X_i}, \widetilde{x_i}) \longrightarrow (X, x_0), \quad i = 1, 2,$$

be covering spaces, and $p_{1*}\pi_1(\widetilde{X_1}, \widetilde{x_1}) \subset p_{2*}\pi_1(\widetilde{X_2}, \widetilde{x_2})$. Then there is a unique covering space $\bar{p} : (\widetilde{X_1}, \widetilde{x_1}) \longrightarrow (X_2, x_2)$ such that $p_2 \circ \bar{p} = p_1$.

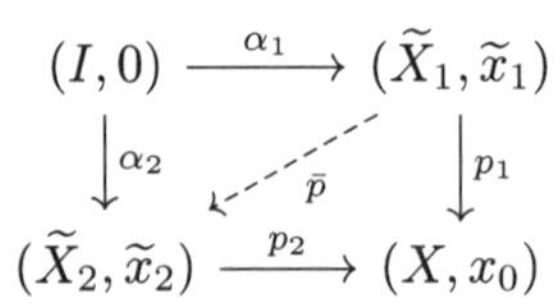

Proof. For each point $\widetilde{x} \in \widetilde{X_1}$ define a path by

$$\alpha_1 : I \longrightarrow \widetilde{X_1} \text{ such that } \alpha_1(0) = \widetilde{x_1}, \alpha_1(1) = \widetilde{x}.$$

Then $p_1 \circ \alpha_1 : I \longrightarrow X is a path$, $p_1(\alpha(0)) = x_0$.

Let $\alpha_2 : I \longrightarrow \widetilde{X_2}$ be the lifting of $p_1 \circ \alpha_1$ such that $\alpha_2(0) = \widetilde{x_2}$. Define a map $\bar{p} : \widetilde{X_1} \longrightarrow \widetilde{X_2}$ by $\bar{p}(\widetilde{x}) = \alpha_2(1)$ for each $\widetilde{x} \in \widetilde{X_1}$.
Then $\bar{p}$ is well-defined: If $\beta_1 : I \longrightarrow \widetilde{X_1}$ is a path such that $\beta_1(0) = \widetilde{x_1}, \beta_1(1) = \widetilde{x}$
then $[\alpha_1 \circ {\beta_1}^{-1}] \in \pi_1(\widetilde{X_1}, \widetilde{x_1})$. Also

$$p_{1*}[\alpha_1 \cdot {\beta_1}^{-1}] \in p_{1*}\pi_1(\widetilde{X_1}, \widetilde{x_1}) \subset p_{2*}\pi_1(\widetilde{X_2}, \widetilde{x_2}),$$

and $p_1 \circ (\alpha_1 \circ \beta_1)$ is a lifting of $\alpha_2 \cdot {\beta_2}^{-1}$, where $p_1 \circ \alpha_1 = p_2 \circ \alpha_2$, $p_1 \circ \beta_1$ $= p_2 \circ \beta_2$. Thus

$$\bar{p}(\widetilde{x}) = \alpha_2(1) = \beta_2(1).$$

Moreover,

$$p_2(\bar{p}(\widetilde{x})) = p_2(\alpha_2(U)) = p_1(\alpha_1(U)) = p_1(\widetilde{x})$$

for every $\widetilde{x} \in \widetilde{X}$. Thus $\bar{p} : \widetilde{X_1} \longrightarrow \widetilde{X_2}$ is a covering space, and $p_2 \circ \bar{p} = p_1$.

□

Definition 9.3.4. Let $p : (\widetilde{X}, \widetilde{x_0}) \longrightarrow (X, x_0)$ be a covering space.

(1) A homeomorphism $h : \widetilde{X} \longrightarrow \widetilde{X}$ such that $p \circ h = p$ is called a **covering transformation** of $f(\widetilde{X}, \widetilde{x_0})$.

(2) $G(\widetilde{X}, p) = \{h : \widetilde{X} \longrightarrow \widetilde{X} \mid h :$ a covering trnsformation of $(\widetilde{X}, \widetilde{x_0})\}$ is called the **covering transformation group** of $(\widetilde{X}, \widetilde{x_0})$.

(3) $(\widetilde{X}, p)$ is a **regular** covering space of X if $p_*\pi_1(\widetilde{X})$ is a normal subgroup of $\pi_1(X)$.

Theorem 9.3.5. *Let X be locally simply connected.*
If $p : (\widetilde{X}, \widetilde{x_0}) \longrightarrow (X, x_0)$ is a regular covering space, then $\pi_1(X, x_0)/p_*\pi_1(\widetilde{X}, \widetilde{x_0})$ and $G(\widetilde{X}, p)$ are isomorphic.

Proof. Let $H := p_*\pi_1(\widetilde{X}, \widetilde{x_0})$. Then $H \lhd \pi_1(X, x_0)$.
Define a map

$$\phi : \pi_1(X, x_0) \longrightarrow G(\widetilde{X}, p)$$

by for each $[\alpha] \in \pi_1(X, x_0)$ and each $[\beta] \in \widetilde{X} = \frac{\Omega(X, x_0)}{H}$

$$\phi([\alpha]) : \widetilde{X} \longrightarrow \widetilde{X}, \phi([\alpha])[\beta] = [\alpha\beta].$$

Then ϕ is well-defined.

Indeed, if $[\alpha_1] = [\alpha_2] \in \pi_1(X, x_0)$ and $\beta_1 \underline{H} \beta_2$, then $[\beta_1\beta_2^{-1}] \in H$, and

$$\begin{aligned}[(\alpha_1\beta_1) \cdot (\alpha_2\beta_2)^{-1}] &= [\alpha_1] \cdot [\beta_1\beta_2^{-1}] \cdot [\alpha_2]^{-1} \\ &= [\alpha_1] \cdot [\beta_1\beta_2^{-1}] \cdot [\alpha_1]^{-1} \in H,\end{aligned}$$

since H is a normal subgroup of $\pi_1(X, x_0)$.
Thus $\alpha_1\beta_1 \underline{H} \alpha_2\beta_2$ and $\phi([\alpha_1])[\beta_1] = [\alpha_1\beta_1] = [\alpha_2\beta_2] = \phi([\alpha_2])[\beta_2]$.

For each $[\alpha] \in \pi_1(X, x_0)$, each $[\beta] \in \widetilde{X}$,

$$\begin{aligned} p(\phi([\alpha]))[\beta] &= p\big[\phi([\alpha])[\beta]\big] = p\big([\alpha\beta]\big) \\ &= (\alpha \cdot \beta)(1) = \beta(1) = p\big([\beta]\big). \end{aligned}$$

So $p \circ \phi([\alpha]) = p$.
Thus $\phi([\alpha]) : \widetilde{X} \longrightarrow \widetilde{X}$ is a map and has inverse $\phi([\alpha^{-1}])$.

For each basic open $([\beta], U)$,

$$\begin{aligned} \phi([\alpha])^{-1}([\beta], U) &= \phi([\alpha^{-1}])([\beta], U) \\ &= ([\alpha^{-1} \cdot \beta], U), \end{aligned}$$

and $\phi([\alpha])$ is continuous.

Similarly, $\phi([\alpha])^{-1} = \phi([\alpha^{-1}])$ is also continuous. Thus

$$\phi([\alpha]) \in G(\widetilde{X}, p).$$

For each $[\alpha_1], [\alpha_2] \in \pi_1(X, x_0)$, $\forall [\beta] \in \widetilde{X}$.

$$\begin{aligned} \phi([\alpha_1][\alpha_2])[\beta] &= \phi([\alpha_1 \cdot \alpha_2])[\beta] = [\alpha_1 \cdot \alpha_2 \cdot \beta] \\ &= \phi([\alpha_1])\phi([\alpha_2])[\beta]. \end{aligned}$$

Thus ϕ is a homomorphism.

Show $\ker \phi = H$:

For each $[\beta] \in \widetilde{X}$,

$$\begin{aligned} [\alpha] \in \ker \phi &\Longleftrightarrow \phi([\alpha_1)[\beta] = [\beta] \Leftrightarrow [\alpha\beta] = [\beta] \\ &\Longleftrightarrow \alpha\beta \underline{H} \beta \Leftrightarrow [\alpha] = [\alpha\beta\beta^{-1}] \in H. \end{aligned}$$

Show $\phi : \pi_1(\widetilde{X}, \widetilde{x_0}) \longrightarrow G(\widetilde{X}, p)$ is onto :
Indeed, if $h \in G(\widetilde{X}, p)$, then $h(x_0) = h([x_0]) = [\alpha]$.
Since $\alpha(1) = p([\alpha]) = p(\widetilde{x_0}) = x_0$, $[\alpha] \in \pi_1(\widetilde{X}, \widetilde{x_0})$. Thus

$$\phi([\alpha]) = h.$$

Because $\phi([\alpha])(x_0) = [\alpha \cdot x_0] = [\alpha] = h([x_0])$, by Theorem 9.2.1, $\phi([\alpha]) = h$. Thus

$$\bar{\phi} : \pi_1(\widetilde{X}, \widetilde{x_0})/p_*\pi_1(\widetilde{X}, \widetilde{x_0}) \longrightarrow G(\widetilde{X}, p)$$

is an isomorphism. □

Corollary 9.3.6. *If* $\pi_1(X, x_0) = \{1\}$, *then* $\pi_1(\widetilde{X}, \widetilde{x_0}) \simeq G(\widetilde{X}, p)$

Example 9.3.7. For universal covering spaces :

$$p_1 : \mathbb{R} \longrightarrow S^1, \quad p_2 : \mathbb{R}^2 \longrightarrow S^1 \times S^1, \quad p_3 : S^2 \longrightarrow \mathbb{RP}^2.$$

We have

$$\begin{aligned}
&\pi_1(S^1) = \mathbb{Z} \simeq G(\mathbb{R}, p_1) \simeq {p_1}^{-1}(\mathrm{pt}), \\
&\pi_1(S^1 \times S^1)\mathbb{Z} \times \mathbb{Z} \simeq G(\mathbb{R}^2, p_2) \simeq {p_2}^{-1}(\mathrm{pt}), \\
&\pi_1(\mathbb{RP}^2) = \mathbb{Z}_2 \simeq G(S^2, p_3) \simeq {p_3}^{-1}(\mathrm{pt}).
\end{aligned}$$

Chapter 10

Differentiable Manifolds

10.1 Differentiable Manifolds

Definition 10.1.1. (A) An n-dimensional Euclidean space $\mathbb{R}^n$ is

(1) $\mathbb{R}^n$ is an **n-dimensional vector space** over $\mathbb{R}$,

(2) there is an **inner product** $< \, , \, >: \mathbb{R}^n \times \mathbb{R}^n \longrightarrow \mathbb{R}$ which is bilinear, symmetric, non-degenerate and positive definite, i.e.,

(i) $< X_1 + X_2, Y_1 + rY_2 >$
$$=< X_1, Y_1 > +r < X_1, Y_2 > + < X_2, Y_1 > +r < X_2, Y_2 >,$$

(ii) $< X, Y >=< Y, X >$,

(iii) if $< X, Y >= 0 \ \ \forall Y,$ then $X = 0$,

(iv) $< X, X >> 0$ if $X \neq 0$.

(B) There is a **metric** $d : \mathbb{R}^n \times \mathbb{R}^n \longrightarrow \mathbb{R}$ on the Euclidean space $\mathbb{R}^n$ given by

$$d(X, Y) =< X - Y, X - Y >^{1/2} .$$

(C) $(\mathbb{R}^n, d)$ is a **metric space**, i.e.,
$U \subset \mathbb{R}^n$ is open iff for any $X \in U$ there is $\varepsilon > 0$ such that

$$B(X, \varepsilon) = \{Y \in \mathbb{R}^n \,|\, d(X,Y) \subset \varepsilon\} \subset U.$$

Definition 10.1.2. A topological space M is a **(topological) n-dimensional manifold** if

(1) M is T_2,

(2) M is second countable,

(3) for each point $x \in M$ there is $U(x)$ a nbd of x in M and n-dimensional open disc $D \subset \mathbb{R}^n$ such that $D \overset{\sim}{\longleftrightarrow} U(x)$ is a homeomorphism.

Example 10.1.3. $\mathbb{R}^0$, $\mathbb{R} = \mathbb{R}^1$, $\mathbb{R}^2$, $\cdots$, $\mathbb{R}^n$, $\cdots$ are manifolds.

(1) $M = \{\text{points}\}$ is a 0-dimensional manifold.

(2) Line, circle, triangle, $\cdots$ are 1-dimensional manifolds.

(3) 2-sphere, torus, box $\cdots$ are 2-dimensional manifolds.

Definition 10.1.4. (1) Let U be open in $\mathbb{R}^n$. A map $f : U \longrightarrow \mathbb{R}^n$ is **smooth** if f has continuous derivatives of all orders on U.

(2) $M \subset \mathbb{R}^n$, for each point $x \in M$ there is U a nbd of x in $\mathbb{R}^n$ and a smooth map $F : U \longrightarrow \mathbb{R}^m$ such that $F = f$ on $U \cap M$. Then we say that the map $f : M \longrightarrow \mathbb{R}^m$ is **smooth**.

(3) Let $M \subset \mathbb{R}^n$ and $N \subset \mathbb{R}^m$. A smooth map $f : M \longrightarrow N$ is bijective and it's inverse $f^{-1} : N \longrightarrow M$ is also smooth. we say that f is a **diffeomorphism**. In this case, we call M and N are **diffeomorphic**.

Example 10.1.5. (1) , 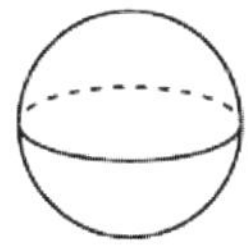, 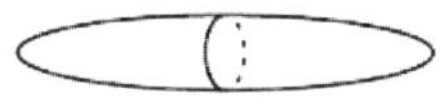 are diffeomorphic.

(2) 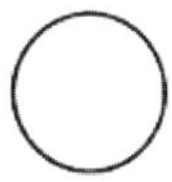, 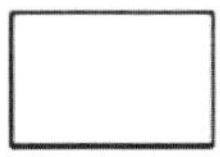, 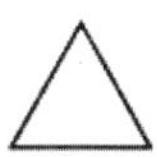 are homeomorphic, not diffeomorphic.

Definition 10.1.6. Let $M \subset \mathbb{R}^n$ be an n-dimensional topological manifold.

(1) M is a **k-dimensional differentiable manifold** if for each point $x \in M$ there is an open set U in $\mathbb{R}^k$, V an open nbd of x in M, and a diffeomophism $\phi : U \longrightarrow V$.

(2) ϕ is called a **parametrization** and ϕ^{-1} is called a **coordinate system**, and denoted by $\phi^{-1} = (x_1, \cdots, x_k)$, where $x_1, \cdots, x_k$ are coordinate functions, $v \in V$, $(x_1(v), \cdots, x_k(v)) \in U$.

(3) $\dim M := k$ is the **dimension** of M.

From now on we assume that all manifolds are differentiable.

Example 10.1.7. The unit sphere

$$S^2 = \{(x_1, x_2, x_3) \in \mathbb{R}^3 \,|\, x_1^2 + x_2^2 + x_3^2 = 1\}$$

is a 2-dimensional manifold.

Proof. Let $U_1 = \{(x_1, x_2) \in \mathbb{R}^2 \,|\, x_1^2 + x_2^2 < 1\}$. Then

$$\phi_1 : U_1 \longrightarrow S^2, \quad \phi_1(x_1, x_2) = (x_1, x_2, \sqrt{1 - (x_1^2 + x_2^2)})$$

is a diffeomorphism to the upper hemi-sphere.
Similarly, we have a diffeomorphism

$$\phi_2 : U_1 \longrightarrow S^2, \quad \phi_2(x_1, x_2) = (x_1, x_2, -\sqrt{1 - (x_1^2 + x_2^2)}).$$

Let $U_2 = \{(x_2, x_3) \in \mathbb{R}^2 \mid x_2^2 + x_3^2 < 1\}$. Then

$$\phi_3 : U_2 \longrightarrow S^2, \quad \phi_3(x_2, x_3) = (\sqrt{1 - (x_2^2 + x_3^2)}, x_2, x_3)$$

$$\phi_4 : U_2 \longrightarrow S^2, \quad \phi_4(x_2, x_3) = (-\sqrt{1 - (x_2^2 + x_3^2)}, x_2, x_3)$$

are diffeomorphisms. Moreover, let $U_3 = \{(x_1, x_3) \in \mathbb{R}^2 \mid x_1^2 + x_3^2 < 1\}$. Then

$$\phi_5 : U_3 \longrightarrow S^2, \quad \phi_5(x_1, x_3) = (x_1, \sqrt{1 - (x_1^2 + x_3^2)}, x_3)$$

$$\phi_6 : U_3 \longrightarrow S^2, \quad \phi_6(x_1, x_3) = (x_1, -\sqrt{1 - (x_1^2 + x_3^2)}, x_3)$$

are diffeomorphisms.
Thus S^2 is a 2-dimensional manifold.

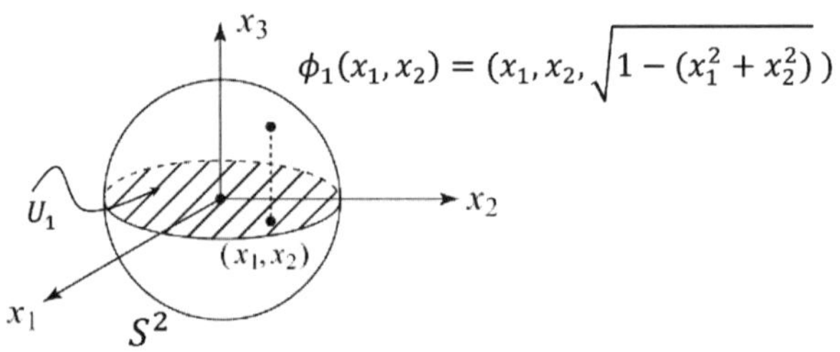

□

Remark 10.1.8. There are many other parametrizations of S^2, for example, the stereographic projections.

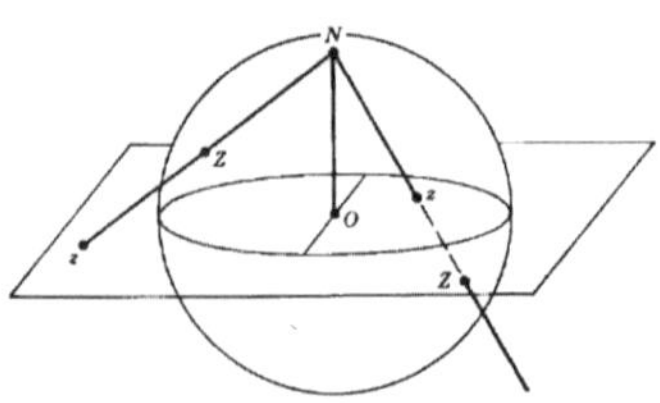

Let $M_1 \subset \mathbb{R}^{n_1}$ be a k_1-dimensional manifold and $M_2 \subset \mathbb{R}^{n_2}$ be a k_2-dimensional manifold. Then for each point $(x_1, x_2) \in M_1 \times M_2$ there are parametrizations ; $U_1 \subset \mathbb{R}^{k_1}$, $U_2 \subset \mathbb{R}^{k_2}$

$$\phi_1 : U_1 \longrightarrow M_1 \quad \text{and} \quad \phi_2 : U_2 \longrightarrow M_2$$

at x_1, x_2, respectively.

Let $\phi := \phi_1 \times \phi_2 : U_1 \times U_2 \longrightarrow M_1 \times M_2$,

$$\phi(y_1, y_2) = (\phi_1(y_1), \phi_2(y_2)).$$

Then $U_1 \times U_2 = \mathbb{R}^{k_1} \times \mathbb{R}^{k_2}$ is open, and ϕ is a parametrization at $(x_1, x_2) \in M_1 \times M_2$.

Thus we have :

Theorem 10.1.9. *If M_1 and M_2 are manifolds, then the product $M_1 \times M_2$ is a manifold with dimension*

$$\dim(M_1 \times M_2) = \dim M_1 + \dim M_2.$$

Let M, $N \subset \mathbb{R}^n$ be manifolds, $N \subset N$. Then N is called a **submanifold** of M.

Example 10.1.10. (1) If $N \subset \mathbb{R}^n$ is a k-dimensional manifold, and N is an open subset of M, then N is a submanifold of dimension k.

(2) For the 2-dimensional manifold S^2, the unit circle

$$S^1 = \{(x_1, x_2, x_3) \in S^2 \mid x_3 = 0\}$$

is a 1-dimensional submanifold of S^2.

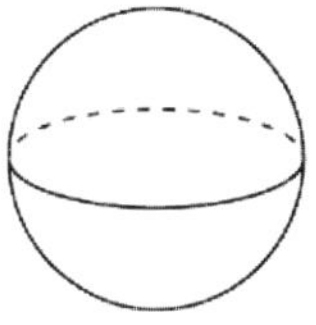

(3) $S^1 \times S^1$ is a 2-dimensional manifold.

(4) $S^n \in \mathbb{R}^{n+1}$,

$$S^n = \{(x_1, \cdots, x_{n+1}) \in \mathbb{R}^{n+1} \mid x_1^2 + \cdots + x_{n+1}^2 = 1\}$$

is an n-dimensional manifold.

(5) $S^1 \times S^2$ is a 3-dimensional manifold.

Example 10.1.11 (Examples of Manifold). (1) Open subsets of $\mathbb{R}^n$ ($\mathbb{C}^n \simeq \mathbb{R}^{2n}$) are n-dimensional (real $2n$-dimension = complex n-dimension) manifolds.

(2) The n-dimensional real projective space :

$$\begin{aligned}\mathbb{RP}^n &= \text{the set of 1-dimensional subspaces of } \mathbb{R}^{n+1}. \\ &= \mathbb{R}^{n+1} - \{0\}/\sim, \quad x \sim y \iff \exists\ \lambda \in \mathbb{R},\ \ y = \lambda x. \\ &= S^n/\sim, \quad x \sim y \iff \text{either } x = y \text{ or } x = -y.\end{aligned}$$

(i) The projection $q : S^n \longrightarrow \mathbb{RP}^n$, $(x_0, \cdots, x_n) \mapsto [x_0, \cdots, x_n]$ defines the quotient topology on $\mathbb{RP}^n$, that is,

$$U \subset \mathbb{RP}^n : \text{ open} \iff \pi^{-1}(U) : \text{ open in } S^n.$$

(ii) If $[x_0, \cdots, x_n] \in \mathbb{RP}^n$, then there exists $i \in \{0, 1, \cdots, n\}$ such that $x_i \neq 0$. Let $U_i := \{[x_0, \cdots, x_n] \in \mathbb{RP}^n \,|\, x_i \neq 0\}$, and

$$\phi_i : U_i \longrightarrow \mathbb{R}^n,$$

$$\phi_i([x_0, \cdots, x_i, \cdots, x_n]) = (\frac{x_0}{x_i}, \cdots, \frac{\widehat{x_i}}{x_i}, \cdots, \frac{x_n}{x_i}).$$

Then ϕ_i is a diffeomorphism and $\{U_i \,|\, i = 0, 1, \cdots, n\}$ is an open cover of $\mathbb{RP}^n$.

(3) The n-dimensional complex projective space :

$$\begin{aligned}\mathbb{CP}^n &= \text{the space of complex 1-dimensional subspaces of } \mathbb{C}^{n+1}. \\ &= \mathbb{C}^{n+1} - \{0\}/\sim, \quad z \sim w \iff \exists\ \lambda \in \mathbb{C},\ \ z = \lambda w. \\ &= S^{2n+1}/\sim, \quad z \sim w \iff z = \lambda w \text{ for some } \lambda \in S^1.\end{aligned}$$

$\mathbb{CP}^n$ is a $2n$-dimensional (n-dimensional complex) manifold.

(4) Let V be a finite dimensional vector space, and $G_k(V)$ the set of k-dimensional subspaces of V.

(i) Let $G_k(\mathbb{R}^n)$ be the set of k-dimensional subspaces of $\mathbb{R}^n$.
Then $G_k(\mathbb{R}^n)$ is a $k(n-k)$-dimensional compact manifold, called a **Grassmannian manifold**.

(ii) Let $G_k(\mathbb{C}^n)$ be the set of k-dimensional complex subspaces of $\mathbb{C}^n$.
Then $G_k(\mathbb{C}^n)$ is a $k(n-k)$-dimensional compact complex manifold, called a **complex Grassmannian manifold**.

(iii) Let $\mathfrak{gl}_{k,n}(\mathbb{R})$ be the set of all $k \times n$-matrices with entries in $\mathbb{R}$.
$\cong \mathbb{R}^{kn}$ which is a kn-dimensional manifold.

(iv) Let Sym be the set of all symmetric $n \times n$-matrices
$= \{A \in \mathfrak{gl}_{n,n}(\mathbb{R}) \mid a_{ij} = a_{ji}\} \simeq \mathbb{R}^{\frac{n(n+1)}{2}}$
is a $\frac{n(n+1)}{2}$-dimensional manifold.

(v) $O(n) = \{A \in \mathfrak{gl}_{n,n}(\mathbb{R}) \mid A^t A = I\}$ is a compact $\frac{n(n-1)}{2}$-dimensional manifold, which has 2 connected components.

(vi) $SO(n) = \{A \in O(n) \mid \det A = 1\}$ is a compact connected $\frac{n(n-1)}{2}$-dimensional manifold.

(5) Let M_1 and M_2 be n-dimensional manifolds, and $x_1 \in M_1$, $x_2 \in M_2$, $D_1(x_1)$ is a small disc with center x_1 in M_1 and $D_2(x_2) \in M_2$. Then

$$\partial D_1(x_1) \simeq S^{n-1} \simeq \partial D_2(x_2).$$

Let

$$M := (M_1 - \mathring{D}_1(x_1)) \bigcup_{S^{n-1}} (M_2 - \mathring{D}_2(x_2))$$

be identified along boundaries.

Then M is an n-dimensional manifold which called the **connected sum** of M_1 and M_2.

(i) Let $T = S^1 \times S^1$ be a torus. The connected sum of g tori is $\sharp_g T$ the **Riemann surface of genus** g.

$T =$ , $\sharp_g T =$ 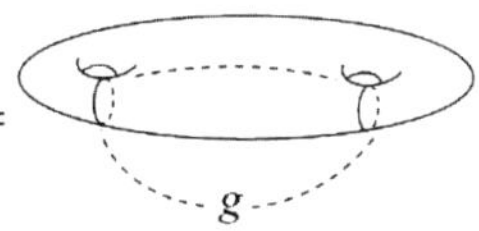

(ii) The connected sum of h projective planes is denoted by $\underset{h}{\sharp}\,\mathbb{RP}^2$.

Theorem 10.1.12 (Classification of compact 2-dimensional manifolds). *Let M be a compact connected 2-dimensional manifold. Then*

(i) *if M is orientable, then either $M \simeq S^2$ or $\underset{g}{\sharp}\,T$ for some g,*

(ii) *if M is nonorientable, then $M \simeq \underset{h}{\sharp}\,\mathbb{RP}^2$ for some h.*

(6) Let $M =$ 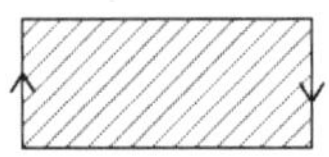 $=$ 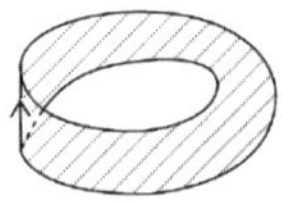 be a Möbius band, and

$\mathbb{RP}^2 =$ 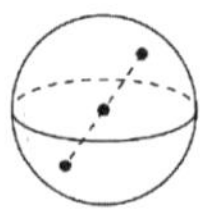 $/x \sim -x =$ 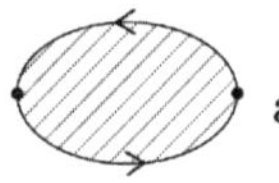 a real projective plane.

Then $\mathbb{RP}^2 \supset \mathbb{RP}^2 - \overset{\circ}{D}{}^2 =$ 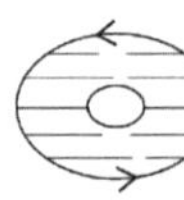 $\overset{(?)}{\simeq}$ 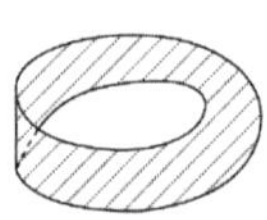 $\simeq M$. Thus

$$\mathbb{RP}^2 \simeq (\mathbb{RP}^2 - \overset{\circ}{D}{}^2)\bigcup_{S^1} D^2 \simeq M \bigcup_{S^1} D^2.$$

(7) The Klein bottle is defined by

$K \simeq$ 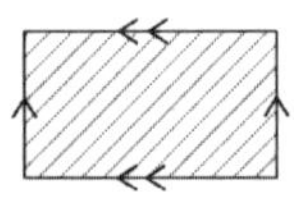 $\overset{(?)}{\simeq} \mathbb{RP}^2 \sharp \mathbb{RP}^2.$

Then $K \sharp \mathbb{RP}^2 \simeq \mathbb{RP}^2 \sharp \mathbb{RP}^2 \sharp \mathbb{RP}^2 \overset{(?)}{\simeq} T \sharp \mathbb{RP}^2$.

(8) $\mathbb{H} = \mathbb{R}^4 =< 1, i, j, k >$ is a quaternion line. Let

$$\mathbb{HP}^n = \text{the set of quaternion line in } \mathbb{H}^{n+1}$$

is a compact $4n$-dimensional manifold, is called the n-dimensional quaternion projective space.

For example $\mathbb{RP}^1 = S^1$, $\mathbb{CP}^1 = S^2$, $\mathbb{HP}^1 = S^4$.

10.1. Exercises

Let all spaces are differentiable manifolds.

1. Prove the map $f : \mathbb{R} \longrightarrow \mathbb{R}$ given by $f(x) = x^3$ is a homeomorphism, but not a diffeomorphism.

2. Let $f : X \longrightarrow Y$ be a smooth map from a manifold of dimension n. Show that

(1) $\mathrm{graph}(f) = \{(x, f(x)) \in X \times Y \,|\, x \in X\}$ is a manifolds of dimension n.

(2) The map $F : X \longrightarrow \mathrm{graph}(f)$, $F(x) = (x, f(x))$ for every $x \in X$, is a diffeomorphism.

3. Let $f_1 : X_1 \longrightarrow Y_1$ and $f_2 : X_2 \longrightarrow Y_2$ be diffeomorphisms. Then show that the product $f : X_1 \times X_2 \longrightarrow Y_1 \times Y_2$, $f(x_1, x_2) = (f_1(x_1), f_2(x_2))$ is also a diffeomorphism.

10.2 Tangent Spaces

Let $U \subset \mathbb{R}^n$ be open, $x \in U$ and $h \in \mathbb{R}^n$, $f : U \longrightarrow \mathbb{R}^m$ a smooth map. Then the derivative of f at x in the direction h is defined by

$$df_x(h) = \lim_{t \to 0} \frac{f(x+th) - f(x)}{t} \quad \in \mathbb{R}^m.$$

Let $(x_1, \cdots, x_n) = x$, $h = (h_1, \cdots, h_n)$ and $f = (f_1, \cdots, f_m)$. Then

$$\begin{aligned} df_x(h) &= \lim_{t\to 0}\Big(\frac{f_1(x_1+th_1,\cdots,x_n+th_n) - f_1(x_1,\cdots,x_n)}{t},\\ &\qquad \cdots, \frac{f_m(x_1+th_1,\cdots,x_n+th_n) - f_m(x_1,\cdots,x_n)}{t}\Big)\\ &= \Big(\frac{\partial f_1}{\partial x_1}(x)h_1 + \cdots + \frac{\partial f_1}{\partial x_n}(x)h_n, \cdots, \frac{\partial f_m}{\partial x_1}(x)h_1 + \cdots\\ &\qquad + \cdots + \frac{\partial f_m}{\partial x_n}(x)h_n\Big)\\ &= \begin{pmatrix} \frac{\partial f_1}{\partial x_1}(x) & \cdots & \frac{\partial f_1}{\partial x_n}(x) \\ \vdots & & \vdots \\ \frac{\partial f_m}{\partial x_1}(x) & \cdots & \frac{\partial f_m}{\partial x_n}(x) \end{pmatrix} \begin{pmatrix} h_1 \\ \vdots \\ h_n \end{pmatrix}. \end{aligned}$$

Theorem 10.2.1 (Chain Rule). *Let $U \subset \mathbb{R}^n$, $V \subset \mathbb{R}^m$ be open, $x \in U$, and maps $f : U \longrightarrow V$, $g : V \longrightarrow \mathbb{R}^\ell$ smooth. Then*

$$d(g \circ f)_x = (dg)_{f(x)} \circ df_x.$$

Remark 10.2.2. (1) If $f : U \longrightarrow \mathbb{R}^m$ is linear, then $df_x : \mathbb{R}^n \longrightarrow \mathbb{R}^m$ is a linear map and $df_x|_U = f$.

(2) df_x is the closest linear map to f near x.

Definition 10.2.3. Let $M \subset \mathbb{R}^n$ be a k-dimensional manifold, and $\phi : U \longrightarrow M$, $\phi(0) = x$, a parametrization. Then $d\phi_0 : \mathbb{R}^k \longrightarrow \mathbb{R}^n$ is a linear map. $\operatorname{Im} d\phi_0 :\equiv T_xM$ is called the **tangent space of M at x.**

Remark 10.2.4.

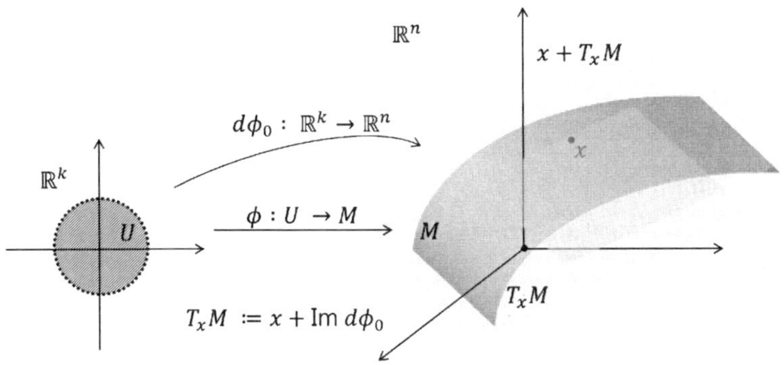

(1) we identify

$$\begin{aligned} T_xM &= \operatorname{Im} d\phi_0 \sim x + \operatorname{Im} d\phi_0 \\ &= x + T_xM, \end{aligned}$$

by parallel translation.

(2) If $x_1 \neq x_2 \in M$, then $T_{x_1}M \cap T_{x_2}M = \emptyset$.

(3) If (x_1, y_1), $(x_2, y_2) \in \mathbb{R}^2$, then $T_{(x_1,y_1)}\mathbb{R}^2 \simeq \mathbb{R}^2 \simeq T_{(x_2,y_2)}\mathbb{R}^2$.
Here $T_{(x_1,y_1)}\mathbb{R}^2 = \{((x_1, y_1), v) \mid v \in \mathbb{R}^2\} \subset \mathbb{R}^4$ and
$T_{(x_2,y_2)}\mathbb{R}^2 = \{((x_2, y_2), w) \mid w \in \mathbb{R}^2\} \subset \mathbb{R}^4$.

Theorem 10.2.5. *If $\phi : U \longrightarrow M$, $\psi : V \longrightarrow M$ are two parameterizations at x, then*

$$d\phi_0\mathbb{R}^k = d\psi_0\mathbb{R}^k = T_xM \subset \mathbb{R}^n.$$

Proof. Suppose that $\phi(0) = \psi(0) = x \in \phi(U) = \psi(V)$. Then the composition $h = \psi^{-1} \circ \phi : U \longrightarrow V$ is a diffeomorphism. Then $\phi = \psi \circ (\psi^{-1} \circ \phi) = \psi \circ h$, and

$$\operatorname{Im} d\phi_0 = \operatorname{Im} d(\psi \circ h)_0 = \operatorname{Im} (d\psi_0 \circ dh_0) \subset \operatorname{Im} d\psi_0.$$

Similarly, considering $\psi = \phi \circ h^{-1}$, we have $\operatorname{Im} d\psi_0 \subset \operatorname{Im} d\phi_0$.
Thus $\operatorname{Im} d\phi_0 = \operatorname{Im} d\psi_0 = T_xM$. □

Theorem 10.2.6. *If M is a k-dimensional manifold and $x \in M$, then the tangent space T_xM has a dimension k.*

Proof. Let $\phi : U \subset \mathbb{R}^k \longrightarrow M \subset \mathbb{R}^n$, $\phi(0) = x$, be a parametrization at x. Then $\phi^{-1} : \phi(U) \subset M \longrightarrow U$ is smooth.

Let $\Phi : W \longrightarrow \mathbb{R}^k$ be an extension of ϕ^{-1} on an open set W in $\mathbb{R}^n$. Then $\Phi \circ \phi = I_U$ and

$$d(\Phi \circ \phi)_0 = d\Phi_x \circ d\phi_0 = I_{\mathbb{R}^k} : \mathbb{R}^k \xrightarrow{d\phi_0} T_xM \xrightarrow{d\Phi_x} \mathbb{R}^k$$

is the identity on $\mathbb{R}^k$.

Thus $d\phi_0 : \mathbb{R}^k \longrightarrow T_xM$ is an isomorphism and $\dim T_xM = k$. □

Let $f : M \longrightarrow N$ be a smooth map between manifolds, $f(x) = y$, $\phi : U \subset \mathbb{R}^k \longrightarrow M$ a parametrization at x and $\psi : V \subset \mathbb{R}^\ell \longrightarrow N$ a parametrization at y, $\phi(0) = x$, $\psi(0) = y$.

Then $h = \psi^{-1} \circ f \circ \phi$ is smooth.

We have the following commutative diagrams :

$$\begin{array}{ccc} M & \xrightarrow{f} & N \\ \uparrow{\scriptstyle \phi} & & \uparrow{\scriptstyle \psi} \\ U & \xrightarrow{h} & V \end{array} \qquad\qquad \begin{array}{ccc} T_xM & \xrightarrow{df_x} & T_yN \\ \uparrow{\scriptstyle d\phi_0} & & \uparrow{\scriptstyle d\psi_0} \\ \mathbb{R}^k & \xrightarrow{dh_0} & \mathbb{R}^\ell \end{array}$$

Definition 10.2.7. We define the **derivative** $df_x : T_xM \longrightarrow T_yN$ of f at x by

$$df_x = d\psi_0 \circ dh_0 \circ d\phi_0^{-1} : T_xM \longrightarrow T_yN.$$

The derivative df_x is independent of the choices ϕ and ψ.

Theorem 10.2.8 (Chain Rule). *Let $M \xrightarrow{f} N \xrightarrow{g} O$ be smooth maps on manifolds. Then for each point $x \in M$*

$$d(g \circ f)_x = dg_{f(x)} \circ df_x.$$

Proof. Let $f(x) = y$, $g(y) = z$, and

$$\phi : U \longrightarrow M \ : \text{ a parametrization at } x, \quad \phi(0) = x,$$

$$\psi : V \longrightarrow N \ : \text{ a parametrization at } y, \quad \psi(0) = y,$$

$$\eta : W \longrightarrow O \ : \text{ a parametrization at } z, \quad \eta(0) = z.$$

Then we have a commutative diagram :

$$\begin{array}{ccccc} M & \xrightarrow{\quad f \quad} & N & \xrightarrow{\quad g \quad} & O \\ \uparrow \phi & & \uparrow \psi & & \uparrow \eta \\ U & \xrightarrow{\quad h \quad} & V & \xrightarrow{\quad k \quad} & W \end{array}$$

where $h = \psi^{-1} \circ f \circ \phi$ and $k = \eta^{-1} \circ g \circ \psi$.

By definition,

$$\begin{aligned} d(g \circ f)_x &= d\eta_0 \circ d(k \circ h)_0 \circ d\phi_0^{-1} \\ &= d\eta_0 \circ dk_0 \circ h_0 \circ d\phi_0^{-1} \quad \text{(Chain rule)} \\ &= d\eta_0 \circ dk_0 \circ d\psi_0^{-1} \circ d\psi_0 \circ dh_0 \circ d\phi_0^{-1} \\ &= dg_y \circ df_x. \end{aligned}$$

Thus $d(g \circ f)_x = dg_{f(x)} \circ df_x$. □

10.2. Exercises

From now on we assume that all spaces are manifolds.

1. Let $f_1 : X_1 \longrightarrow Y_1$ and $f_2 : X_2 \longrightarrow Y_2$ be smooth maps. Then prove the followings :

(1) $T_{(x_1,x_2)}(X_1 \times X_2) = T_{x_1}(X_1) \times T_{x_2}(X_2)$.

(2) $d(f_1 \times f_2)_{(x_1,x_2)} = d(f_1)_{x_1} \times d(f_2)_{x_2}$.

(3) Let $F_1 : X_1 \longrightarrow X_1 \times Y_1$ be $F_1(x_1) = (x_1, f_1(x))$. Then

$$d(F_1)_{x_1}(v) = (v, d(f_1)_{x_1}(v))$$

for each $x_1 \in X_1$ and $v \in T_{x_1}(X_1)$.

(4) Let $p_1 : X_1 \times X_2 \longrightarrow X_1$ be $p_1(x_1, x_2) = x_1$. Then

$$d(p_1)_{(x_1,x_2)}(v_1, v_2) = v_1.$$

(5) Let $I_1 : X_1 \longrightarrow X_1 \times X_2$ be $I_1(x_1) = (x_1, x_2)$ where x_2 a fixed point in X_2. Then $d(I_1)(v_1) = (v_1, 0)$.

2. Let $\triangle$: $X \longrightarrow X \times X$ be the diagonal, $\triangle\ (x) = (x, x)$ for every $x \in X$. Show

(1) $d(\triangle)_x(v) = (v, v)$.

(2) The map $h : T(X) \longrightarrow N(\triangle, X \times X)$

$$h(x, v) = ((x, x), (v, -v))$$

is a diffeomorphism, where $N(\triangle, X \times X)$ is called the normal bundle of $\triangle$ in $X \times X$.

10.3 Immersion and Imbedding

Let M and N be manifolds with same dimension. If $f : M \to N$ is a local diffeomorphism at x (i.e., $\exists\ U$: a nbd of x, V: a nbd of $y = f(x)$ the restriction $f : U \longrightarrow V$ is a diffeomorphism), then the derivative $df_x : T_xM \longrightarrow T_yN$ is an isomorphism. Since

$$f^{-1}f = I_U, \quad ff^{-1} = I_V, \quad df^{-1} \circ df = I$$

and

$$df \circ df^{-1} = I.$$

Theorem 10.3.1 (Inverse function theorem). *If $df_x : T_xM \longrightarrow T_yN$ is a linear isomorphism, then $f : M \to N$ is local diffeomorphism at x.*

Proof. Consider parametrizations :

$$\begin{array}{ccc} M & \xrightarrow{\ f\ } & N \\ {\scriptstyle\phi}\uparrow & & \uparrow{\scriptstyle\psi} \\ U & \xrightarrow{\ h\ } & V \end{array}$$

where $h = \psi^{-1} \circ f \circ \phi$, $\phi(0) = x$, $\psi(0) = y = f(x)$. The derivative $dh_0 = d\psi_0^{-1} \circ df_x \circ d\phi_0$ is an isomorphism.

By inverse function theorem in Euclidean space, h is a local diffeomorphism at 0.

Thus $f = \psi \circ h \circ \phi^{-1}$ is a local diffeomorphism at x. $\square$

Remark 10.3.2. (1) By inverse function theorem,

$f : M \longrightarrow N$ is a local diffeomorphism at x

iff $df_x : T_xM \longrightarrow T_{f(x)}N$ is an isomorphism

iff $\det(df_x) \neq 0$.

(2) $f : M \longrightarrow N$ is a local diffeomorphism at every point

$\not\Longrightarrow f : M \longrightarrow N$ is a diffeomorphism.

Example 10.3.3. $f : \mathbb{R} \longrightarrow S^1$, $f(x) = e^{2\pi i x}$, is a local diffeomorphism at every point, but not a diffeomorphism, because $df_x = 2\pi i e^{2\pi i x}$.

(3) For inverse function theorem,

$$\dim M = \dim N.$$

Let us assume $\dim M \leq \dim N$.

Definition 10.3.4. (1) If $df_x : T_x M \longrightarrow T_{f(x)N}$ is injective, then f is called an **immersion** at x.

(2) If $f : M \longrightarrow N$ is an immersion at every point of M, then f is called an **immersion**.

(3) For $k \leq \ell$, the inclusion

$$i : \mathbb{R}^k \longrightarrow \mathbb{R}^\ell, \quad i(x_1, \cdots, x_k) = (x_1, \cdots, x_k, 0, \cdots, 0)$$

is an immersion. The i is called the **standard immersion**.

Theorem 10.3.5 (Local Immersion Theorem). *Let $f : M \longrightarrow N$ be an immersion at x, $y = f(x)$. Then there are parametrizations such that*

$$\phi : U \longrightarrow M, \quad \phi(0) = x,$$
$$\psi : V \longrightarrow N, \quad \psi(0) = y,$$

satisfying $f(x_1, \cdots, x_k) = (x_1, \cdots, x_k, 0, \cdots, 0)$.

Proof. Consider parametrizations :

$$\begin{array}{ccc} M & \xrightarrow{\quad f \quad} & N \\ \phi\Big\uparrow & & \psi\Big\uparrow \\ 0 \in U \subset \mathbb{R}^k & \xrightarrow{\quad g \quad} & V \subset \mathbb{R}^\ell \end{array}$$

Since df_x is injective and $df_x \circ d\phi_0 = d\psi_0 \circ dg_0$, dg_0 is injective.

By a change of basis of $\mathbb{R}^\ell$, we may assume $dg_0 = \begin{pmatrix} I_k \\ 0 \end{pmatrix}$ is a $\ell \times k$-matrix, here I_k is the $k \times k$-unit matrix.

Define $G : U \times \mathbb{R}^{\ell-k} \longrightarrow \mathbb{R}^\ell$ by $G(x,z) = g(x)+(0,z)$. Then $dG_0 = I_\ell$ is a $(\ell \times \ell)$-unit matrix.

By inverse function theorem, G is a local diffeomorphism at 0 in $U \times \mathbb{R}^{\ell-k} \subset \mathbb{R}^\ell$. Let $G : V' \longrightarrow V$ be a diffeomorphism. Then $\psi \circ G : V' \longrightarrow N$ is a parametrization at y. Considering the following diagram:

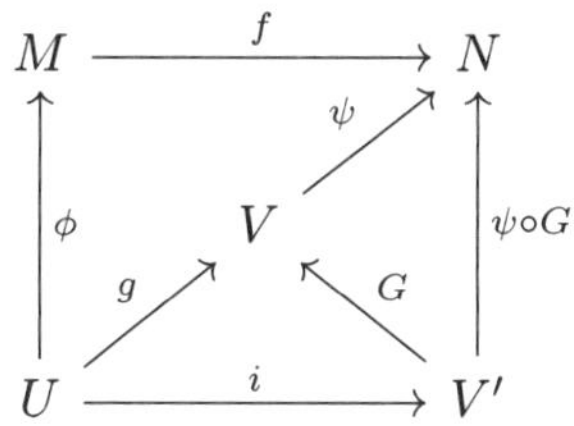

Thus $f \circ \phi = \psi \circ g = \psi \circ (G \circ i) = (\psi \circ G) \circ i$ and

$$f(x_1, \cdots, x_k) = (x_1, \cdots, x_k, 0, \cdots, 0).$$

□

Remark 10.3.6. (1) If $f : M \longrightarrow N$ is immersion set x, then f is an immersion on a small nbd of x.

(2) If $\dim M = \dim N$, then $f : M \longrightarrow N$ is immersion at x, and f is a local diffeomorphism at x.

(3) If $f : M \longrightarrow N$ is a local diffeomorphism and bijection, then f is a diffeomorphism.

(4) Let $f : M \longrightarrow N$ be an immersion. Then each point $x \in M$ has a nbd W such that $f : W \longrightarrow f(W) \subset N$ is a diffeomorphism. Each point of $f(M)$ is a coordinate system of N. However, $f(W) \subset f(M)$ may not be open in $f(M)$ and $f(M)$ may not be a submanifold of N.

Example 10.3.7. (1) Let $f : S^1 \longrightarrow 8$ be an immersion, surjective, not injective. Then $f(S^1) = 8$ is not a manifold.

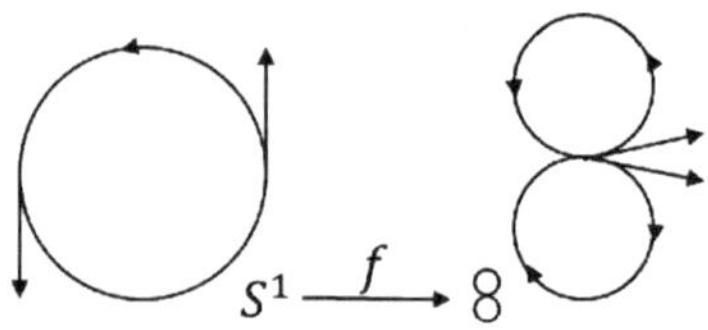

(2) There is a map $f : \mathbb{R}^1 \longrightarrow \mathbb{R}^2$, which is an immersion, $f : \mathbb{R}^1 \longrightarrow f(\mathbb{R}^1)$ and bijective, but $f(\mathbb{R}^1)$ is not a manifold.

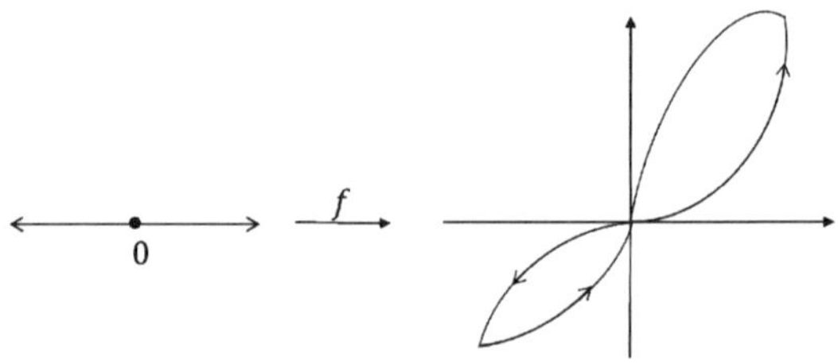

(3) Let $f : \mathbb{R}^2 \longrightarrow T = S^1 \times S^2$ be defined by $f(s,t) = (e^{2\pi is}, e^{2\pi it})$. If $L = \{(x, ax) \,|\, x \in \mathbb{R}\}$ is a line in $\mathbb{R}^2$, where a is an irrational, then $f : L \longrightarrow f(L)$ is immersion, bijection, and $f(L)$ is dense in T. But $f(L)$ is not a submanifold of T.

Definition 10.3.8. (1) A map $f : M \longrightarrow N$ is **proper** if the inverse image of every compact subset of N is compact in M. i.e., If C is compact, then $f^{-1}(C)$ is compact.

(2) If $f : M \longrightarrow N$ is immersion, injective and proper, then f is called an **embedding**.

Theorem 10.3.9. *If $f : M \longrightarrow N$ is an embedding, then $f(M)$ is a submanifold of N.*

Proof. Let $W \subset M$ be a coordinate nbd at $x \in M$. Show $f(W)$ is open in $f(M)$. $f(M)$ is a manifold. In fact, suppose that there is $y \in f(W)$ which is not an interior.
Thus there is a sequence $< y_i >$, $y_i \in f(M) - f(W)$ such that $< y_i >$ converges to y.

The set $\{y\} \cup \{y_i \,|\, i = 1, 2, \cdots\}$ is compact in $f(M) \subset N$.
Since f is proper, $\{f^{-1}(y)\} \cup \{f^{-1}(y_i) \,|\, i = 1, \cdots\}$ is compact in M. There is subsequence $< f^{-1}(y_i) >$ such that $f^{-1}(y_i) \longrightarrow x_0 \in M$. The sequence $y_i \longrightarrow f(x_0)$. Since $y_i \longrightarrow y$, $f(x_0) = y$. So

$$x_0 \in f^{-1}(cy) \subset f^{-1}(f(W)) = W.$$

Since $f(y_i) \longrightarrow x_0$ and $x_0 \in W$ open. There is $n_0 > 0$ such that $f^{-1}(y_i) \in W$ for all $i > n_0$.
Thus $y_i \in f(W)$ for all $i > n_0$. This contradicts the construction of $< y_i >$. □

10.3. Exercises

Prove or solve the followings:

1. Let $h : \mathbb{R}^n \longrightarrow \mathbb{R}^n$ be a linear map and $v \in \mathbb{R}^n$. Then the map $f(x) = h(x) + v$ is a diffeomorphism iff the determinant $\det(h) \neq 0$.

2. Let $f : X \longrightarrow Y$ and $g : Y \longrightarrow Z$ be smooth.

(1) If f and g are immersions, then $g \circ f$ is.

(2) If $\dim X = \dim Y$, then f is an immersion iff f is a local diffeomorphism.

3. Let n be a positive integer, and $f : \mathbb{R} \longrightarrow \mathbb{R}$, $f(t) = nt$, and $g : \mathbb{R} \longrightarrow S^1$, $g(t) = e^{2\pi i t}$, and $h : S^1 \longrightarrow S^1$, $h(z) = z^n$. Then

(1) f, g, and h are immersions.

(2) $g \circ f = h \circ g$ and is an immermsion.

10.4 Submersion

Let $f : M \longrightarrow N$ be a smooth map.

Then there are three cases :

(1) $\dim M = \dim N$: local diffeomorphism, diffeomorphism.

(2) $\dim M \leq \dim N$: immersion, embedding.

(3) $\dim M \geq \dim N$: submersion, implicit function theorem.

Definition 10.4.1. (1) $f : M \longrightarrow N$ is a **submersion at** $x \in M$ if $df_x : T_xM \to T_{f(x)}N$ is surjective.

(2) $f : M \longrightarrow N$ is a **submersion** if f is a submersion at every point of M.

(3) For $k \geq \ell$, the projection

$$p : \mathbb{R}^k \longrightarrow \mathbb{R}^\ell, \quad p(x_1, \cdots, x_\ell, \cdots, x_k) = (x_1, \cdots, x_\ell)$$

is the **standard submersion**.

Theorem 10.4.2 (Local submersion Theorem)**.** *Let $f : M \longrightarrow N$ be a submersion at x, $f(x) = y$. Then there are parametrizations*

$$\phi : U \subset \mathbb{R}^k \longrightarrow M, \quad \phi(0) = x,$$
$$\psi : V \subset \mathbb{R}^\ell \longrightarrow N, \quad \psi(0) = y,$$

such that $f(x_1, \cdots, x_\ell, \cdots, x_k) = (x_1, \cdots, x_\ell)$.

Proof. Consider parametrizations :

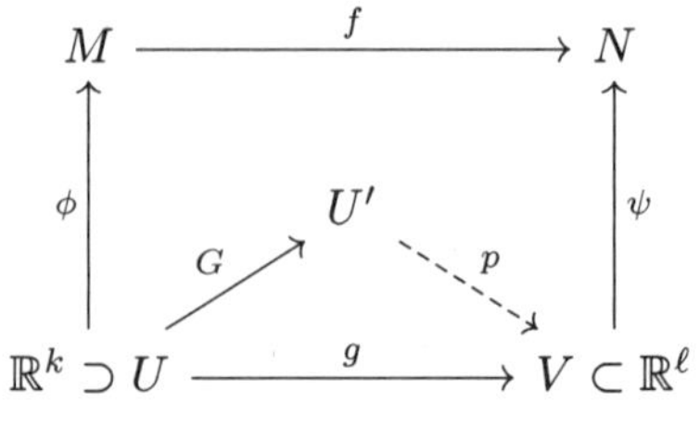

$$
\begin{aligned}
f(x) = y, &\quad \phi(0) = x,\\
\psi(0) = y, &\quad g(0) = 0,\\
f \circ \phi &= \psi \circ g.
\end{aligned}
$$

Then $df_x \circ d\phi_0 = d\psi_0 \circ dg_0$, since df_x is onto, dg_0 is onto.

By a coordinate change, we may assume that $dg_0 = (I_\ell, 0)$ is a $(\ell \times k)$-matrix. Define $G : U \subset \mathbb{R}^k \longrightarrow \mathbb{R}^k$ by

$$G(a_1, \cdots, a_\ell, \cdots, a_k) = (g(a_1, \cdots, a_k), a_{\ell+1}, \cdots, a_k).$$

Then $dG_0 = I_k$ is the $(k \times k)$-unit matrix and $G(0) = 0$. Thus G is a local diffeomorphism at 0.

Say $G : U \longrightarrow U'$ a local diffeomorphism and $p : U' \longrightarrow V$ the standard submersion. Since $f \circ \phi = \psi \circ g$ and $g = p \circ G$,

$$\phi \circ G^{-1} : U' \longrightarrow M$$

is a parametrization at x, and

$$f \circ (\phi \circ G^{-1}) = \psi \circ (g \circ G^{-1}) = \psi \circ p.$$

Thus $f(x_1, \cdots, x_\ell \cdots, x_k) = (x_1, \cdots, x_\ell)$. $\square$

Remark 10.4.3. If $f : M \longrightarrow N$ is a submersion at x, then f is a submersion on a small nbd of x.

Indeed, by Theorem 10.4.2, there is coordinate systems at x and $y = f(x)$ such that

$$(x_1, \cdots, x_\ell, \cdots, x_k) = (x_1, \cdots, x_\ell).$$

Then $df_x = (I_\ell, 0)$. Considering the following diagram :

$$
\begin{array}{ccc}
M & \xrightarrow{\ A\ } & \{((df_x)_{i,j}) \mid i, j = 1, \cdots, \ell\} \\
 & {\scriptstyle \det \circ A} \searrow & \downarrow {\scriptstyle \det} \\
 & & \mathbb{R}
\end{array}
$$

$A(x) = (df_x)_{\ell\times\ell}$.

The composition det $\circ$ A is continuous and

$$\Big(\det \circ A\Big)(x) = \det(A(x)) = \det(I_{(\ell\times\ell)}) = 1.$$

Let $U = \Big(\det \circ A\Big)^{-1}(\frac{1}{2}, \frac{3}{2})$. Then $x \in U$, U open, and f is a submersion on U.

Definition 10.4.4. Let $f : M \longrightarrow N$ be a smooth map.

(1) $y \in N$ is a **regular vale of** f if $x \in f^{-1}(y)$, then df_x is surjective.

(2) $y \in N$ is a **critical value of** f if y is not a regular value of f.

If $y \in N$ is a regular vale of $f : M \longrightarrow N$, then for each $x \in f^{-1}(y)$. There are coordinate systems $U(x)$ and $V(y)$ such that

$$f : U(x) \longrightarrow V(y), \quad f(x_1, \cdots, x_\ell, \cdots, x_k) = (x_1, \cdots, x_\ell).$$

Thus $f^{-1}|_U(y) = f^{-1}|_U(0) = U(x) \cap f^{-1}(0)$
$= \{(0, \cdots, 0, x_{\ell+1}, \cdots, {}_k) \in U\}$
is a $(k-\ell)$-dimensional open disc in $U(x)$.

Theorem 10.4.5 (Preimage Theorem). *If $y \in N$ is a regular value of $f : M \longrightarrow N$ and $f^{-1}(y) \neq \emptyset$, then $f^{-1}(y)$ is a submanifold of M with dimension*

$$\dim f^{-1}(y) = \dim M - \dim N.$$

Remark 10.4.6. Let $f : M \longrightarrow N$ be a smooth map.

(1) If $y \in N - f(M)$, then y is a regular value.

(2) Let $\dim M > \dim N$.
Then $y \in N$ is a regular value of f iff for any $x \in f^{-1}(y)$ df_x is onto.

(3) Let dim$M$ = N.
Then y is a regular value of f iff for any $x \in f^{-1}(y)$ a smooth function f is a local diffeomorphism at x.

(4) Let dim$M$ < N. Then y is a regular value of f iff $y \in N - f(M)$.

Example 10.4.7. (1) Let $f : \mathbb{R}^k \longrightarrow \mathbb{R}$ be the map defined by

$$f(x_1, x_2, \cdots, x_k) = x_1^2 + \cdots + x_k^2, \quad k \geq 1.$$

Then $f^{-1}(1) = S^{k-1} = \{(x_1, \cdots, x_k) \in \mathbb{R}^k \,|\, x_1^2 + \cdots + x_k^2 = 1\}$ is a $(k-1)$-dimensional manifold.

Indeed, for $\forall x \in f^{-1}(1) = S^{k-1}$, the derivative $df_x : \mathbb{R}^k \longrightarrow \mathbb{R}$ given by

$$df_x = (2x_1, \cdots, 2x_k) \neq (0, \cdots, 0)$$

is onto. Thus 1 s a regular value of f.

(2) Orthogonal Group

$$O(n) = \{A : \mathbb{R}^n \longrightarrow \mathbb{R}^n \,|\, AA^t = I\}$$

is $\frac{n(n-1)}{2}$-dimensional manifold.

Indeed, let $\mathfrak{gl}(n) = \{A \,|\, A : \ (n \times n)\text{-matrix}\}$
$= \{A : \mathbb{R}^n \longrightarrow \mathbb{R}^n : \text{ linear}\}$
$\approx \mathbb{R}^{n^2}$ be the n-dimensional linear group.
$O(n) = \{A \in \mathfrak{gl}(n) \,|\, AA^t = I\}$ is the orthogonal group.
$\mathrm{Sym}(n) = \{A \in \mathfrak{gl}(n) \,|\, A = A^t\}$ is the symmetric matrices,

$$\dim \mathrm{Sym}(n) = \frac{n(n+1)}{2}.$$

Define a map $f : \mathfrak{gl}(n) \longrightarrow \mathrm{Sym}(n)$ by $f(A) = AA^t \in \mathrm{Sym}(n)$, since $(AA^t)^t = AA^t$.

Clearly, f is smooth and $O(n) = f^{-1}(I)$. Show that I is a regular value of f. For each $A \in O(n)$, show $df_A : \mathfrak{gl}(n) \longrightarrow \mathrm{Sym}(n)$ is onto.

For any $C \in \mathrm{Sym}(n)$, $C = C^t$,

$$\begin{aligned} df_A(\frac{1}{2}(A)) &= \lim_{h\to 0} \frac{1}{h}\Big[f(A + \frac{1}{2}hCA) - f(A)\Big] \\ &= \lim_{h\to 0} \frac{1}{n}\Big[(A + \frac{1}{2}hCA)(A + \frac{1}{2}hCA)^t - AA^t\Big] \\ &= \lim_{h\to 0} \frac{1}{n}\Big[(A + \frac{1}{2}hCA)(A^t + \frac{1}{2}hA^tC^t) - AA^t\Big] \end{aligned}$$

$$\begin{aligned} &= \lim_{h\to 0} \frac{1}{n}\Big[AA^t + \frac{1}{2}hCAA^t + \frac{1}{2}hAA^tC^t \\ &\qquad + \frac{1}{4}h^2CAA^tC^t - AA^t\Big] \\ &= \lim_{h\to 0} \Big[\frac{1}{2}C + \frac{1}{2}C^t + \frac{1}{4}hCC^t\Big] \\ &= C, \end{aligned}$$

since $C = C^t$ and $AA^t = I$.

Thus I is a regular value of f, and $O(n) = f^{-1}(I)$ is a smooth manifold with

$$\begin{aligned} \dim O(n) &= \dim \mathfrak{gl}(n) - \dim \mathrm{Sym}(n) \\ &= n^2 - \frac{n(n+1)}{2} = \frac{n(n-1)}{2}. \end{aligned}$$

□

Let $k \geq \ell$, and a k-dimensional manifold M. Let $g_1, \cdots, g_\ell : M \longrightarrow \mathbb{R}$,

$$g = (g_1, \cdots, g_\ell) : M \longrightarrow \mathbb{R}^\ell, \quad g(x) = (g_0(x), \cdots, g_\ell(x)).$$

If $0 \in \mathbb{R}^\ell$ is a regular value of g, then $g^{-1}(0) = g_1^{-1}(0) \cap \cdots \cap g_\ell^{-1}(0)$ is a $(k - \ell)$-dimensional submanifold of M.

If $dg_x : T_xM \longrightarrow \mathbb{R}^\ell$ is onto

$$\Big(\text{ i.e., for each } i = 1, \cdots, \ell,\ (d(g_i)_x : T_xM \longrightarrow \mathbb{R}\text{: onto }\Big),$$

then $g_1, \cdots, g_\ell$ are called **independent** at x.

Corollary 10.4.8. *Suppose $k \geq \ell$ and M is a k-dimensional manifold. If $g_1, \cdots, g_\ell : M \longrightarrow \mathbb{R}$ are independent on $Z = \cap_{i=1}^{\ell} g_i^{-1}(0)$, then Z is a submanifold of M with dimension* $\dim Z = k - \ell$.

Definition 10.4.9. Let M be a k-dimensional manifold and $Z \subset M$ a submanifold of $\dim(k-\ell)$. Then $\dim M - \dim Z := \operatorname{codim} Z = \ell$ is called the **codimension** of Z in M.

Theorem 10.4.10. *If $y \in N$ is a regular value of $f : M \longrightarrow N$, then $f^{-1}(y)$ is the common zero of* $(\dim N) = \ell$ *independent functions* $M \longrightarrow \mathbb{R}$.

Proof. Let $\phi : W \subset N \longrightarrow \mathbb{R}^\ell$ be a coordinate system at $y \in N$, $\phi(y) = 0$. The inverse $f^{-1}(W)$ is open in M and $f^{-1}(y) \subset f^{-1}(W)$.

Let $g = \phi \circ f : f^{-1}(W) \subset M \longrightarrow W \longrightarrow \mathbb{R}^\ell$.

Then $g^{-1}(0) = f^{-1}(\phi^{-1}(0)) = f^{-1}(y)$ and

$$\forall x \in g^{-1}(0), \quad dg_x = d\phi_{f(x)} \circ df_x \quad \text{is onto.}$$

Thus $f^{-1}(y) = g^{-1}(0) = \cap_{i=1}^{\ell} g_i^{-1}(0), \quad g = (g_1, \cdots, g_\ell)$ □

Theorem 10.4.11. *Every submanifold of M is locally a common zero of independent functions on M.*

Proof. Let Z be a submanifold of M with $\operatorname{codim} Z = \ell$. For each point $x \in Z \subset M$, by Theorem 10.3.5, there is a parametrization $\phi : U \subset \mathbb{R}^k \longrightarrow W(x) \subset M$ such that $(x_1, \cdots, x_{k-\ell}, 0, 0, \cdots, 0)$: a coordinated systems on $W(x) \cap Z$.

Thus $W(x) \cap Z = \cap_{i=1}^{\ell} (\phi^{-1})_{k-\ell+i}(0)$ where $\phi^{-1} = (\phi_1^{-1}, \cdots, \phi_k^{-1})$. □

Theorem 10.4.12. *Let $y \in N$ be a regular value of a map $f : M \longrightarrow N$, and $Z = f^{-1}(y)$. Then for any $x \in Z$, $df_x : T_M \longrightarrow T_yN$ has the kernel $df_x^{-1}(0) = T_xZ$.*

Proof. Since $y \in N$ is a regular value of $f : M \longrightarrow N$, $Z = f^{-1}(y)$ is a submanifold of M with dimension $\dim Z = \dim M - \dim N$. For each $x \in Z$, $df_x : T_xM \longrightarrow T_yN$. By definition $df_x(T_xZ) = 0$ and $T_xZ \subset \operatorname{Ker} df_x$.

$$\dim df_x^{-1}(0) = \dim T_xM - \dim T_yN = \dim T_xZ.$$

Thus $df_x^{-1}(0) = T_xZ$. □

10.4. Exercises

1. Let $f : X \longrightarrow Y$ be a smooth map.

(1) If f is a submersion, then show that f is open.

(2) If X is compact, Y connected, and f a submersion, then f is surjective.

(3) If X is compact, $\dim X = \dim Y$, and y is a regular value of f, then show that

  (i) $f^{-1}(y)$ is a finite set $\{x_1, \cdots, x_n\}$ in X.

  (ii) there exists a neighborhood V of y in Y, such that $f^{-1}(V) = \cup_{i=1}^{n} U_i$ is a disjoint union and the restriction $f : U_i \longrightarrow V$ is a diffeomorphism for each $i = 1, \cdots, n$, where U_i is a neighborhood of x_i in X.

2. Prove that the projection map

$$p : T(X) \longrightarrow X, \quad p(x, v) = x,$$

(1) is a submersion.

(2) $d(p)_{(x,v)} : T_{(x,v)}(TX) \longrightarrow T_xX$ has the vertical tangent space $\{0\} \times T_xX$ as its kernel.

10.5 Transversality

If a map $f : M \longrightarrow N$ has a regular value $y \in N$, then $f^{-1}(y)$ is a submanifold of M and

$$\dim M - \dim f^{-1}(y) = \dim N - \dim\{y\}.$$

Definition 10.5.1. Let $Z \subset N$ be a submanifold of N. A map $f : M \longrightarrow N$ is **transversal to** Z, $f \pitchfork Z$, if for every point $x \in f^{-1}(Z)$ $\mathrm{Im}(df_x) + T_{f(x)}Z = T_{f(x)}N$, where $\dim(df_x) + T_{f(x)}Z$ is the vector space generated by $\mathrm{Im}(df_x)$ and $T_{f(x)}Z$, may not

$$\mathrm{Im}(df_x) \bigcap T_{f(x)}Z = \{0\}.$$

Theorem 10.5.2. *If a map $f : M \longrightarrow N$ is transversal to a submanifold Z in N, then $f^{-1}(Z)$ is a submanifold of M with*

$$\mathrm{codim} f^{-1}(Z) = \dim M - \dim f^{-1}(N) = \dim N - \dim Z = \mathrm{codim} Z.$$

Proof. For every $x \in f^{-1}(Z)$, the point $y = f(x) \in Z$ has a coordinate neighborhood V in N such that $V \cap Z$ is the common zero of $\ell = \dim N - \dim Z$ independent functions $g_1, \cdots, g_\ell : V \longrightarrow \mathbb{R}$. Then there is a coordinate neighborhood W of x in M, such that

$$g_1 \circ f, \cdots, g_\ell \circ f : W \longrightarrow V \longrightarrow \mathbb{R}$$

independent at x and $W \cap f^{-1}(Z)$ is the common zero of $g_1 \circ f$, $\cdots$, $g_\ell \circ f$.

Let $g = (g_1, \cdots, g_\ell) : V \longrightarrow \mathbb{R}^\ell$. Then $g \circ f : W \longrightarrow V \longrightarrow \mathbb{R}^\ell$ has its derivative $d(g \circ f)_x = dg_{f(x)} \circ df_x$. Since $\mathrm{Ker}\,(dg_{f(x)}) = T_{f(x)}Z$ and $dg_{f(x)}$ is onto, $g \circ f$ is submersion at x iff

$$\mathrm{Im}\, df_x + T_{f(x)}Z = T_{f(x)}N.$$

Since f is transversal to Z, $g \circ f$ is a submersion on $f^{-1}(Z) \cap W$, and $g_1 \circ f, \cdots, g_\ell \circ f$ are independent.
Thus $f^{-1}(Z)$ is a submanifold with dimension

$$\begin{aligned}\dim f^{-1}(Z) &= \dim M - \ell \\ &= \dim M - \dim N + \dim Z\end{aligned}$$

i.e., $\dim M - \dim f^{-1}(Z) = \dim N - \dim Z$. □

Remark 10.5.3. If a map $f : M \longrightarrow N$ is transversal to $Z = \{y\} \subset N$, then $T_yZ = \{0\}$ and for every $x \in f^{-1}(y)$, $df_x(T_xM) + T_yZ = T_yN$. So $df_x(T_xM) = T_yN \quad \forall x \in f^{-1}(y)$. The y is a regular value of f.

Example 10.5.4. (1) Let $f : \mathbb{R}^1 \longrightarrow \mathbb{R}^2$ be a map defined $f(x) = (0, x)$.

① If $Z = x$-axis, then $f \pitchfork Z$.

② If $Z = y$-axis, then $f \not\pitchfork Z$.

③ If $Z = \{(x, \sin x) \mid x \in \mathbb{R}\}$, then $f \pitchfork Z$.
If $g(x) = (x, 0)$, then $g \pitchfork Z$.

(2) Let $g : \mathbb{R} \longrightarrow \mathbb{R}^2$, $g(x) = (x, x^2)$. Then

$$g \pitchfork y\text{-axis} := Z_1 \quad \text{and} \quad g \not\pitchfork x\text{-axis} := Z_2.$$

Since $g^{-1}(Z_1) = \{0\}$ and $dg_0 = \begin{bmatrix} 1 \\ 0 \end{bmatrix}$, $\operatorname{Im} dg_0 = x$-axis, $T_0Z_1 = y$-axis.
Then $g^{-1}(Z_2) = \{0\}$ and $\operatorname{Im} dg_0 = x$-axis $= T_0Z_2$.

Let M, Z be submanifolds of N, and $i : M \longrightarrow N$ inclusion. Then $i^{-1}(Z) = M \cap Z$. For every point $x \in M \cap Z$, $di_x : T_xM \longrightarrow T_xM$ is inclusion. Thus $i \pitchfork Z$ iff

$$\forall x \in M \cap Z \quad T_xM + T_xZ = T_xN.$$

In this case denote $M \pitchfork Z$ we say that **M is transversal to Z** in N.

Theorem 10.5.5. *Let M and Z be submanifolds of N. If $M \pitchfork Z$, then $i^{-1}(Z) = M \cap Z$ is a submanifold of N and*

$$\operatorname{codim}(M \cap Z) = \operatorname{codim} M + \operatorname{codim} Z.$$

Remark 10.5.6. (1) Transversality of submanifolds M and Z does depend on N.

Example. If M = x-axis, Z = y-axis, then $M \pitchfork Z$ in $\mathbb{R}^2$ but $M \not\pitchfork Z$ in $\mathbb{R}^3$.

(2) If $\dim M + \dim Z < \dim N$, then

$$M \pitchfork Z \iff M \cap Z = \emptyset.$$

(3) If $M = \mathbb{R}^2$ is the xy-plane, $Z = S^2$ and $N = \mathbb{R}^3$, then $M \pitchfork Z$ and $M \cap Z = S^1$.

10.5. Exercises

1. Let $f : X \longrightarrow Y$, $g : Y \longrightarrow Z$ be smooth, Z_1 a submanifold of Z, and $Y_1 = g^{-1}(Z_1)$.

(1) Show that if $g \pitchfork Z_1$ and $y_1 \in Y_1$, $g(y_1) = z_1$, then

$$T_{y_1}(Y_1) = dg_{x_1}^{-1}[T_{z_1}(Z_1)]$$

(2) Show that if $g \pitchfork Z_1$, then $f \pitchfork Y_1$ iff $g \circ f \pitchfork Z_1$.

2. Let X_1 and X_2 be submanifolds of $X$and $x \in X_1 \cap X_2$. Show that if $X_1 \pitchfork X_2$, then

$$T_x(X_1 \cap X_2) = T_x(X_1) \cap T_x(X_2).$$

3. Let $\triangle$ be the diagonal of $\mathbb{R}^n \times \mathbb{R}^n$ and $A : \mathbb{R}^n \longrightarrow \mathbb{R}^n$ a linear map. Show that graph$(A) \pitchfork \triangle$ iff $+1$ is not an eigenvalue of A.

10.6 Sard Theorem and Morse Function

Theorem 10.6.1 (Sard Theorem). *Let $f : X \longrightarrow Y$ be a smooth map between manifolds. Then almost every point of Y is a regular value of Y where almost every (a.e.) point of Y is a subset Y which is a measure zero set.*

Remark 10.6.2. (1) Let X be a k-dimensional manifold. A subset $C \subset X$ has **measure zero** if and only if for every parametrization $\phi : U \subset \mathbb{R}^k \longrightarrow X$, the inverse image $\phi^{-1}(C)$ has measure 0 in $\mathbb{R}^k$.

(2) If $A \subset \mathbb{R}^k$ has measure 0, and $f : \mathbb{R}^k \longrightarrow \mathbb{R}^k$ smooth, then $f(A)$ also has measure 0.

Corollary 10.6.3. (1) *The regular values of a smooth map $f : X \longrightarrow Y$ is dense in Y.*

(2) *Let $f_i : X_i \longrightarrow Y$ be a countable family of smooth maps. Then the set of regular values for all f_i is dense in Y.*

Proof. (1) By Sard Theorem, the set of critical values of f has measure 0. The regular values is dense in Y.

(2) Let $C_i \subset Y$ be the set of critical values of f_i. For any $\varepsilon > 0$ and for each i, there is $\{I^i_j \,|\, j = 1, \cdots\}$ of k-dimensional rectangles such that

$$C_i \subset \bigcup_j I^i_j \quad \text{and} \quad \sum_j \operatorname{vol}(I^i_j) < \frac{\varepsilon}{2^i}.$$

The set $\{I^i_j\}_{i,j}$ covers $\bigcup_i C_i$ and $\sum_{i,j} \operatorname{vol}(I^i_j) < \sum_i \frac{\varepsilon}{2^i} = \varepsilon$. Thus $\bigcup_i C_i$ has measure 0.

By (1) the set of regular values for all f_i is dense in Y.

□

Definition 10.6.4. For a smooth map $f : X \longrightarrow Y$,

(1) $x \in X$ is a **regular point** of f if df_x is surjective.

(2) $x \in X$ is a **critical point** of f if x is not a regular point of f.

Remark 10.6.5. For a smooth map $f : X \longrightarrow Y$,

(1) the regular points and critical points are in X, but the regular values and critical values are in Y.

(2) the set of critical values has measure 0, the set of critical points may not have measure 0.

Let X be a manifold, and $f : X \longrightarrow \mathbb{R}$ smooth. Then $df_x : T_xX \longrightarrow \mathbb{R}$ is either 0 or nonzero.

If $df_x \neq 0$, then x is a regular point of f. There is a coordinate neighborhood U of x such that

$$f : U \longrightarrow \mathbb{R}, \quad f(x_1, \cdots, x_k) = x_1$$

is the projection to the first coordinate.

If $df_x = 0$, then x is a critical point of f. In this case f may have either local maximum or local minimum or saddle point at x.

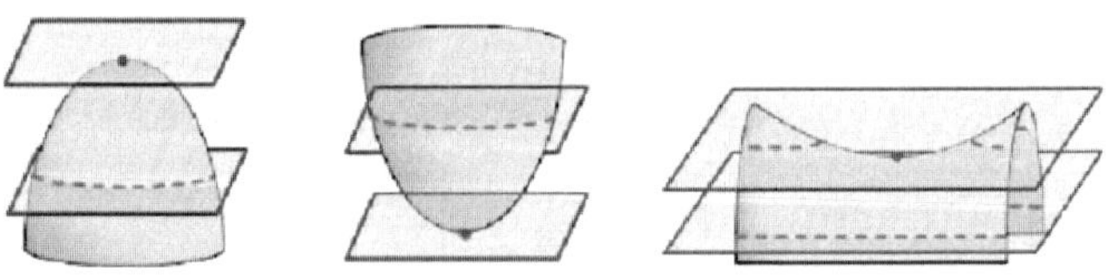

Thus a critical point x gives some geometric information of X.

Let $f : \mathbb{R}^k \longrightarrow \mathbb{R}$ have a critical point x. Then $df_x = 0$, that is,

$$\frac{\partial f}{\partial x_1}(x) = 0, \ \cdots, \ \frac{\partial f}{\partial x_k}(x) = 0.$$

Definition 10.6.6. (1) Let $H(f)(x) = (\frac{\partial^2 f}{\partial x_i \partial x_j}(x))$ be a $(k \times k)$-matrix. Then $H(f)(x)$ is called **Hessian matrix** of f at x.

(2) A point x is called a **nondegenerate critical point** of f
if (i) x is a critical point, i.e., $df_x = 0$,
(ii) $H(f)(x)$ is nonsingular, i.e., $\det H(f)(x) \neq 0$.

Theorem 10.6.7. *The nondegenerate critical points of a smooth map $f : \mathbb{R}^k \longrightarrow \mathbb{R}$ are isolated.*

Proof. Let $g : \mathbb{R}^k \longrightarrow \mathbb{R}^k$ be the map defined by

$$g = \Big(\frac{\partial f}{\partial x_1}, \cdots, \frac{\partial f}{\partial x_k}\Big).$$

Then $df_x = 0$ iff $g(x) = 0$.

The derivative $dg_x = (\frac{\partial^2 f}{\partial x_i \partial x_j}) = H(f)(x)$,
dg_x is an isomorphism iff $\det(dg_x) = \det[H(f)(x)] \neq 0$
iff x is a nondegenerate critical point of f
iff there are $U(x)$ a nbd of x and $V(0)$ a nbd of 0 in $\mathbb{R}^k$,
such that $g : U(x) \longrightarrow V(0)$ is a diffeomorphism.

In this case $g(x') \neq 0$ for all $x' \in U(x) - \{x\}$. Thus the nondegenerate critical point x of f is isolated. □

Theorem 10.6.8 (Morse Theorem). *If $a \in \mathbb{R}^k$ is a nondegenerate critical point of a map $f : \mathbb{R}^k \longrightarrow \mathbb{R}$, then there is a local coordinate system $(x_1, \cdots, x_k)$ at a such that $f(x) = f(a) + \sum_{i=1}^k \varepsilon_i x_i^2, \quad \varepsilon_i = \pm 1$.*

Let a smooth map $f : X \longrightarrow \mathbb{R}$ have a critical point x and $\phi : U \longrightarrow X$ a parametrization at x, $\phi(0) = x$. Then

$$d(f \circ \phi)_0 = df_x \circ, d\phi_0 = 0,$$

and 0 is a critical point of $f \circ g$. A point $x \in X$ is a **nondegenerate critical point** of f if 0 is a nondegenerate critical point of $f \circ \phi$.

Lemma 10.6.9. *If* 0 *is a nondegenerate critical point of* $f : \mathbb{R}^k \longrightarrow \mathbb{R}$, $\psi : \mathbb{R}^k \longrightarrow \mathbb{R}^k$, $\psi(0) = 0$, *a diffeomorphism, then* 0 *is a nondegenerate critical point of* $f \circ \psi$.

Proof. Let $f_1 = f \circ \psi$, H Hessian of f, and H_1 Hessian of $f \circ \psi$. By chain rule,

$$\frac{\partial f_1}{\partial x_i}(x) = \sum_\alpha \frac{\partial f}{\partial x_\alpha}(\psi(x))\frac{\partial \psi_\alpha}{\partial x_i}(x).$$

$$\frac{\partial^2 f_1}{\partial x_i \partial x_j}(0) = \sum_\alpha \sum_\beta \frac{\partial^2 f}{\partial x_\alpha \partial x_\beta}(0)\frac{\partial \psi_\alpha}{\partial x_i}(0)\frac{\partial \psi_\beta}{\partial x_j}(0).$$

Since 0 is a critical point of f, $d(f_1)_0 = df_0 \circ d\psi_0 = 0$,

$$H_1 = (d\psi_0)^t H (d\psi_0).$$

Since ψ is a diffeomorphism, $\det(d\psi_0) \neq 0$. $\det H_1 = (\det \psi_0)^2 \det H$. Thus $\det H \neq 0$ iff $\det H_1 \neq 0$. □

Definition 10.6.10. A map $a : X \longrightarrow \mathbb{R}$ is a **Morse function** if every critical point of f is nondegenerate.

Let $X \subset \mathbb{R}^n$ be a manifold and $f : X \longrightarrow \mathbb{R}$ a smooth map. $a = (a_1, a_2, \cdots, a_n) \in \mathbb{R}^n$. Define $f_a : X \longrightarrow \mathbb{R}$ by

$$f_a(x) = f(x) + a_1x_1 + \cdots + a_nx_n.$$

Theorem 10.6.11. *Let* $f : X \longrightarrow \mathbb{R}$ *be a smooth map. Then for almost every point* $a \in \mathbb{R}^n$ f_a *is a Morse function.*

Lemma 10.6.12. *Let* $U \subset \mathbb{R}^k$ *be open,* $f : U \longrightarrow \mathbb{R}$ *a smooth map. Then for almost every* $a \in \mathbb{R}^k$, $f_a(x) = f(x) + a_1x_1 + \cdots + a_kx_k$ *is Morse.*

Proof. Let $g : U \longrightarrow \mathbb{R}^k$, $g = \left(\frac{\partial f}{\partial x_1}, \cdots, \frac{\partial f}{\partial x_k}\right)$. Then the derivative at p is

$$\begin{aligned}(df_a)_p &= \left(\frac{\partial f_a}{\partial x_1}(p), \cdots, \frac{\partial f_a}{\partial x_k}(p)\right)\\ &= \left(\frac{\partial f}{\partial x_1}(p) + a_1, \cdots, \frac{\partial f}{\partial x_k}(p) + a_k\right)\\ &= g(p) + a.\end{aligned}$$

Thus p is a critical point of f_a iff $g(p) = -a$, $H(f)_p = H(f_a)_p = dg_p$. If $-a$ is a regular value of g and $g(p) = -a$, then dg_p is a linear isomorphism.

If $-a$ is a regular value of g, every critical point of f_a is nondegenerate. By Sard theorem, a.e. $a \in \mathbb{R}^k$, $-a$ is a regular value of g. □

Proof of 10.6.11. Let $x = (x_1, \cdots, x_n) \in X \subset \mathbb{R}^n$, and $(x_1, \cdots, x_k)$ be a coordinate system of x in X. Let $\{U_\alpha\}$ be a countable open cover of X such that for each α, $(x_1, \cdots, x_k)$ a coordinate system U_α, and

$$f_{(0,c)} = f + c_{k+1}x_{k+1} + \cdots + c_n x_n, \quad c = (c_{k+1}, \cdots, c_n).$$

By Lemma 10.6.12 for a.e. $b \in \mathbb{R}^k$

$$f_{(b,c)} = f_{(0,c)} + b_1 x_1 + \cdots + b_k x_k$$

is a Morse function on U_α.

Let $S_\alpha = \{a \in \mathbb{R}^n \mid f_a \text{ is not Morse on } U_\alpha\}$. Then $S_\alpha \cap (\mathbb{R}^k \times \{c\})$ has measure 0 in $\mathbb{R}^k$. By Funibi theorem S_α has measure 0 in $\mathbb{R}^n$. Thus $\{a \in \mathbb{R}^n \mid f_a \neq \text{Morse}\} = \bigcup_\alpha S_\alpha$ has measure zero, since $\{U_\alpha\}_\alpha$ is a countable open cover of X.

10.6. Exercises

1. Let S^n be the unit sphere in $\mathbb{R}^{n+1}$ and $h : S^n \longrightarrow \mathbb{R}$, $h(x_1, \cdots, x_{n+1}) = x_{n+1}$, the height map. Show that h is a Morse map with two critical points.

2. Let X be a compact manifold in $\mathbb{R}^n$, $f : X \longrightarrow \mathbb{R}$ a map, and for each point $a = (a_1, \cdots, a_n) \in \mathbb{R}^n$

$$f_a = f + a_1x_1 + \cdots + a_nx_n.$$

Show that

(1) $X_1 = \{a \in \mathbb{R}^n \mid f_a \text{ is Morse}\}$ is an open subset in $\mathbb{R}^n$.

(2) If X is not compact, then

$$X_2 = \{a \in \mathbb{R}^n \mid f_a : \text{ Morese }\}$$

is a countable intersection of open subsets of $\mathbb{R}^n$.

3. Let X be a manifold of n-dimension and

$$S(X) = \{(x, v) \in T(X) \mid \ |v| = 1\}.$$

Show that :

(1) If $g : T(X) \longrightarrow \mathbb{R}$ is defined $g(x, v) = v \cdot v$, then find the set of regular values of g.

(2) Show that $S(X)$ is a $(2n-1)$-dimensional submanifold of $T(X)$.

10.7 Manifolds with Boundary

(1) $D^n = \{x \in \mathbb{R}^n \mid |x| \leq 1\}$ has the boundary $\partial D^n = S^{n-1}$.

(2) $S^1 \times [0,1]$ has the boundary $\partial(S^1 \times [0,1]) = S^1 \times \{0,1\}$

(3) $\mathbb{H}^k = \{(x_1, \cdots, x_k) \in \mathbb{R}^k \mid x_k \geq 0\}$ has its boundary

$$\mathbb{R}^{k-1} \simeq \mathbb{R}^{k-1} - \{0\} = \partial\mathbb{H}^k \subset \mathbb{R}^k.$$

$D^n =$ 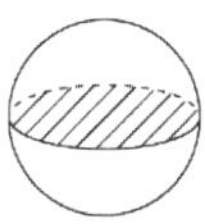, $S^1 \times [0,1] =$ 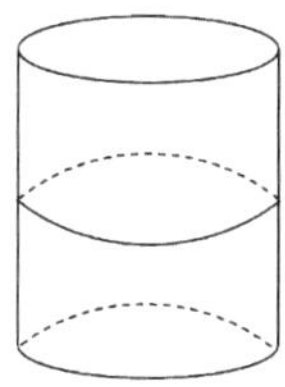, $\mathbb{H}^k =$ 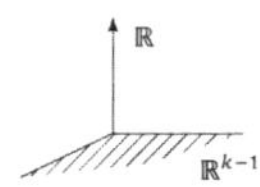

Definition 10.7.1. $X \subset \mathbb{R}^n$ is a **k-dimensional manifold with boundary** if for each point $x \in X$ there is an open nbd $V(x) \subset X$ of x which is homeomorphic to an open set in $\mathbb{H}^k$.
i.e., $\phi : U(p) \subset \mathbb{H}^k \longrightarrow V(x) \subset X$ is a homeomorphism.

$$\partial X = \{x \in X \mid \phi,: U(0) \longrightarrow V(x) \text{ is a parametrization, } \phi(0) = x\}$$

is called the **boundary** of X, and $i(X) = X - \partial X$ the **interior** of X.

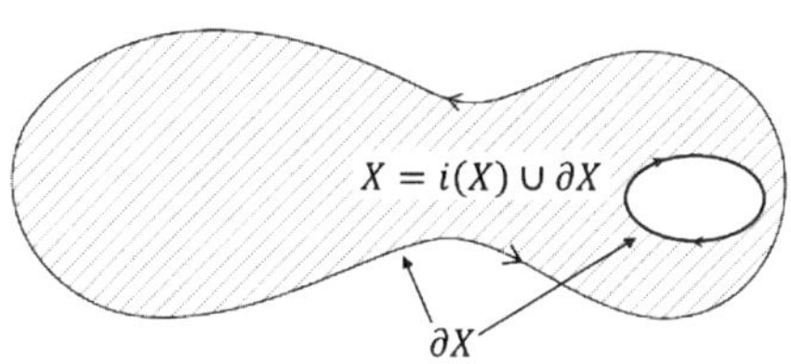

Remark 10.7.2. (1) Let $X \subset \mathbb{R}^n$ be a k-dimensional manifold with boundary. Then $i(X)$ is the interior of X in X, but may be not an interior in $\mathbb{R}^n$. The $i(X)$ is a k-dimensional manifold without boundary. And the boundary ∂X is a $(k-1)$-dimensional manifold without boundary.

(2) The rectangle $X = [0,1] \times [0,1] =$ 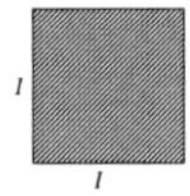 $\underset{\text{homeo. not diffeo.}}{\simeq}$  is not a manifold with boundary, since he vertices $(0,0)$, $(0,1)$, $(1,0)$, $(1,1)$ have no smooth parametrizations.

Theorem 10.7.3. *Let X be a manifold without boundary, and Y a manifold with boundary. Then the product $X \times Y$ is a manifold with boundary*

$$\partial(X \times Y) = X \times \partial Y$$

and

$$\dim(X \times Y) = \dim X + \dim Y.$$

Proof. Let $U \subset \mathbb{R}^k$, $V \subset \mathbb{H}^\ell$ be open. $\phi : U \longrightarrow X$ and $\psi : V \longrightarrow Y$ are parametrizations. Then $U \times V \subset \mathbb{R}^k \times \mathbb{H}^\ell = \mathbb{H}^{k+\ell}$ and

$$\phi \times \psi : U \times V \longrightarrow X \times Y$$

is a parametrization. □

Let $g : U \subset \mathbb{H}^k \longrightarrow \mathbb{R}^\ell$ be a smooth map and $u \in \partial U$. Then

$$dg_u : \mathbb{R}^k \longrightarrow \mathbb{R}^\ell$$

is defined as follows :

Choose $\widetilde{U}$is an open nbd of u in $\mathbb{R}^k$, and an extension $\widetilde{g} : \widetilde{U} \longrightarrow \mathbb{R}^\ell$ of g, define $dg_u := d\widetilde{g}_u : \mathbb{R}^k \longrightarrow \mathbb{R}^\ell$. Then dg_u is well-defined.

Indeed, let $\widetilde{\widetilde{g}} : \widetilde{\widetilde{U}} \longrightarrow \mathbb{R}^\ell$ be another extension of g. Choose a sequence $u_n \in i(U)$ such that $u_n \longrightarrow u$. Then on $i(U)$ $g = \widetilde{g} = \widetilde{\widetilde{g}}$ and $d\widetilde{g}_{u_i} = d\widetilde{\widetilde{g}}_{u_i}$ for all i. Taking limits

$$\lim_{i \longrightarrow \infty} d\widetilde{g}_{u_i} = \lim_{i \to \infty} d\widetilde{\widetilde{g}}_{u_i} = d\widetilde{g}_u = d\widetilde{\widetilde{g}}_u = dg_u.$$

Let X be a k-dimensional manifold with boundary in $\mathbb{R}^n$. Then for $x \in \partial X$, $T_x X$ is a k-dimensional subspace of $\mathbb{R}^n$, and $T_x \partial X$ is a $(k-1)$-dimensional subspace of $T_x X$.

Theorem 10.7.4. *Let X be a k-dimensional manifold with boundary ∂X. Then ∂X is a $(k-1)$-dimensional manifold without boundary.*

Proof. Let

$x \in \partial X = \{x \in X \mid \phi : U(0) \subset \mathbb{H}^k \longrightarrow V(x) \subset X$ is a parametrization$\}$.

Since $\partial U(0) = U(0) \cap \mathbb{R}^{k-1}$ is an open nbd of 0 in $\mathbb{R}^{k-1}$,
$\partial V(x) = V(x) \cap \partial X$ is an open nbd of x in ∂X.
Show $\phi(\partial U(0)) = \partial V(x)$. Then

$$\phi|_{\partial U(0)} : \partial U(0) \subset \mathbb{R}^{k-1} \longrightarrow \partial V(x) \subset \partial X$$

is a diffeomorphism. By definition, $\phi(\partial U(0)) \subset \partial V(x)$.

Show $\phi(\partial U(0)) \supset \partial V(x)$. If $v \in \partial V(x) \subset V(x)$, then there is $u \in U(0)$ such that $\phi(u) = r$. Assume $u \in i(U(0))$. Then there is a parametrization

$$\psi : W(0) \subset \mathbb{H}^k \longrightarrow V_1(v) \subset X,$$

$\psi(0) = v$, $V_1(v) \subset V(x)$. Then

$$g = \phi^{-1} \circ \psi : W(0) \longrightarrow V_1(v) \longrightarrow [\text{a nbd of } u]$$

in $U(0)$ is a smooth map and $g(0) = \phi^{-1} \circ \psi(0) = \phi^{-1}(v) = u$.
Thus dg_0 is an isomorphism. By inverse function theorem, g is a local diffeomorphism.
Contradicts! Thus $u \in \partial U(0)$ and $v = \phi(u) \in \phi(\partial U(0))$. □

Remark 10.7.5. (1) Theorem 10.7.4, if $x \in \partial X \subset X$, then $T_x \partial X$ is a $(k-1)$-dimensional sub-vector space of $T_x X$.

(2) For a map $f : X \longrightarrow Y$, we denote the restriction of f by $\partial f : \partial X \longrightarrow Y$. The derivative is

$$d(\partial f)_x = df_x|_{T_x(\partial X)}.$$

Theorem 10.7.6. *Let X be a manifold with boundary, $Z \subset Y$ manifolds without boundary, and $f : X \longrightarrow Y$ smooth, $f \pitchfork Z$ and $\partial f : \partial X \longrightarrow Y$ smooth, $\partial f \pitchfork Z$. Then*

(1) *$f^{-1}(Z)$ is a manifold with boundary*

$$\partial(f^{-1}(Z)) = f^{-1}(Z) \cap \partial X,$$

(2) $\operatorname{codim} f^{-1}(Z) = \operatorname{codim} Z$.

Proof. By Theorem 10.5.2, since $f \pitchfork Z$ and

$$f|_{i(X)} \pitchfork Z, \quad f^{-1}(Z) \cap i(X) = f|_{i(X)}^{-1}(Z)$$

is a manifold satisfying codimension condition.

For each point $x \in \partial f^{-1}(Z) = f^{-1}(Z) \cap \partial X$ and $\ell = \operatorname{codim} Z$ in Y, let ϕ be a submersion from a coordinate nbd V of $f(x) \in Y$ to $\mathbb{R}^\ell$. Then $Z \cap V = \phi^{-1}(0)$. Let U be a coordinate nbd of x in X such that $f(U) \subset V$. Then $\phi \circ f : U \longrightarrow V \longrightarrow \mathbb{R}^\ell$ is a submersion and $(\phi \circ f)^{-1}(0) = f^{-1}(Z) \cap U$.

Let $h : W \subset \mathrm{H}^k \longrightarrow U \subset X$ be a parametrization. Then

$$g = \phi \circ f \circ h : W \longrightarrow \mathbb{R}^\ell$$

is a smooth map. Since $h : W \longrightarrow U$ is a diffeomorphism,

$$\begin{aligned} &f^{-1}(Z) \cap U \text{ is a manifold with boundary} \\ \text{iff } &(f \circ h)^{-1}(Z) = g^{-1}(0) \end{aligned}$$

is a manifold with boundary at $u = h^{-1}(x) \in \partial W$. Since $f \pitchfork Z$,

$$\begin{aligned} &df_x(T_xX) + T_{f(x)}Z = T_{f(x)}Y, \\ &\text{iff } x \text{ is a regular point of } \phi \circ f \\ &\text{iff } u \text{ is a regular point of } g \end{aligned}$$

Since g is smooth there is a nbd $\widetilde{W} \subset \mathbb{R}^k$ of u and an extension $\widetilde{g}$ on $\widetilde{W}$ of g such that $d\widetilde{g}_u = dg_u$ and u is a regular point of $\widetilde{g}$. Thus $\widetilde{g}^{-1}(0) := S \subset \mathbb{R}^k$ is a manifold without boundary and

$$g^{-1}(0) = S \cap \mathbb{H}^k = \{s \in S \mid \pi(s) \geq 0\}$$

where $\pi : S \in \mathbb{R}^k \longrightarrow \mathbb{R}$ is the projection to the last coordinate.

Show 0 is a regular value of $\pi : S \longrightarrow \mathbb{R}$. If 0 is not a regular value of π, there is $s \in S$, $\pi(s) = 0$, such that $d\pi_s = 0$. Thus $s \in S \cap \partial\mathbb{H}^k$ and $\pi : \mathbb{R}^k \longrightarrow \mathbb{R}$ is a projection and linear. And $d\pi_s = \pi$, $d\pi_s(T_sS) = 0$, and $T_sS \subset T_s(\partial\mathbb{H}^k) = \mathbb{R}^{k-1}$. Since $s \in \widetilde{g}^{-1}(0)$ and $dg_s = d\widetilde{g}_s : \mathbb{R}^k \longrightarrow \mathbb{R}$ and $T_sS = \operatorname{Ker} dg_s$.

$$d(\partial g)_s : \mathbb{R}^{k-1} \longrightarrow \mathbb{R}^\ell \text{ and } \operatorname{Ker} d(\partial g)_s = \operatorname{Ker} dg_s = T_sS.$$

Since dg_s and $d(\partial g)_s$ are onto (since $\partial f \pitchfork Z$, $\partial g \pitchfork 0$)

$$\dim(\operatorname{Ker} dg_s) = k - \ell, \quad \dim(\operatorname{Ker} d(\partial g)_s) = k - \ell - 1.$$

Contradict! Thus 0 is a regular value of π. □

Lemma 10.7.7. *Let S be a manifold without boundary, and $\pi : S \longrightarrow \mathbb{R}$, 0 regular value of π. Then $\{s \in S \mid \pi(s) \geq 0\}$ is a manifold with boundary $\pi^{-1}(0)$.*

Proof. Since $\pi^{-1}((\infty, 0))$ is an open subset of S, $\pi^{-1}(\infty, 0)$ is a manifold with dimension $\dim S$. Since 0 is a regular value of π, $\pi^{-1}(0)$ is a submanifold of s with dimension $\dim S - 1$.
Moreover π is the standard submersion at $s \in \pi^{-1}(0)$. □

Example 10.7.8. Let $S = \mathbb{R}^n$, $\pi : \mathbb{R}^n \longrightarrow \mathbb{R}$, $\pi(s) = 1 - |s|^2$. Then $\pi^{-1}[0, \infty) = \{s \in \mathbb{R}^n | \, |s| \leq 1\}$ is an n-dimensional unit disc with boundary $\pi^{-1}(0) = \{s \in \mathbb{R}^n \mid |s|^2 = 1\} = S^{n-1}$ which is an $(n-1)$-dimensional sphere.

Theorem 10.7.9 (Sard Theorem). *Let X be a manifold with boundary, Y a manifold without boundary, and $f : X \longrightarrow Y$ be a smooth map. Then almost every point of Y is a regular value of f and $\partial f : \partial X \longrightarrow Y$.*

Proof. For each point $x \in \partial X \subset X$, since $T_x\partial X \subset T_x X$, if x is a regular point of ∂f, then x is a regular point of f. If $y \in Y$ is a critical value of $f : i(X) \longrightarrow Y$ or $\partial f : \partial X \longrightarrow Y$, then y is a critical value of f or ∂f.

Since $i(X)$ and ∂X are manifolds without boundary, the sets of the critical values of $f : i(X) \longrightarrow Y$ and $\partial f : \partial X \longrightarrow Y$ have measure zero, respectively.

The union of two measure zero sets is also measure 0.
Thus almost every point of Y is a regular values of f and ∂f. □

10.7. Exercises

1. Let $f : X \longrightarrow Y$ be a diffeomorphism between manifolds with boundaries. Show that the restriction map $\partial f : \partial X \longrightarrow \partial Y$ is a diffeomorphism.

2. Let X be a manifold with boundary. Show that

(1) ∂X is a closed subset of X.

(2) if X is compact, then ∂X is.

(3) if ∂X is compact, then X may not be compact.

3. Show that the solid hyperbolid

$$X = \{(x_1, \cdots, x_n) \in \mathbb{R}^n \,|\, x_1^2 + \cdots + x_{n-1}^2 - x_n^2 \leq 1\}$$

is a manifold with boundary of dimension n.

10.8 One Dimensional Manifolds

Theorem 10.8.1. *Compact connected 1-dimensional manifolds are diffeomorphic to either the unit interval* $[0,1]$ *or the unit circle* S^1.

Theorem 10.8.2. *A compact 1-dimensional manifold has an even number of boundary points.*

Theorem 10.8.3. *Let* X *be a compact manifold with boundary. Then there is no smooth map* $f : X \longrightarrow \partial X$ *such that* $f|_{\partial X} = I_{\partial X}$.

Proof. Suppose that $f : X \longrightarrow \partial X$ is a smooth map $\partial f = I_{\partial X}$. If $x \in \partial X$ is a regular value of f, then $f^{-1}(x)$ is a compact 1-dimensional manifold. Thus

$$\partial(f^{-1}(x)) = f^{-1}(x) \bigcap \partial X = \{x\}.$$

This contradicts to $\sharp\partial(f^{-1}(x))$ which is even. □

Theorem 10.8.4 (Brouwer Fixed Point Theorem). *Let*

$$B^n = \{x \in \mathbb{R}^n \mid \ |x| \le 1\}$$

be the unit disc in $\mathbb{R}^n$ *Then any smooth map* $f : B^n \longrightarrow B^n$ *has a fixed point (i.e., there a point* $x \in B^n$ *such that* $f(x) = x$*).*

Proof. Suppose that $f : B^n \longrightarrow B^n$ is a smooth map such that $f(x) \neq x$ for every $x \in B^n$. Define a map $g : B^n \longrightarrow \partial B^n$ by

$$g(x) = \overrightarrow{f(x)x} \bigcap \partial B^n$$

then $\partial g = I_{\partial B^n}$.

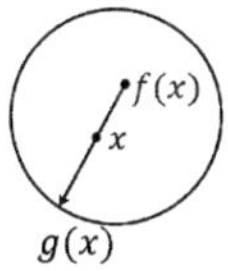

i.e., $g(x) = tx + (1-t)f(x)$, for some $t \geq 1$. Then

$$1 = g(x) \cdot g(x) = t^2|x - f(x)|^2 + 2tf(x) \cdot (x - f(x)) + |f(x)|^2,$$

and

$$t^2|x - f(x)|^2 + 2tf(x)(x - f(x)) + |f(x)|^2 - 1 = 0.$$

Thus

$$t = \frac{-f(x)(x - f(x))}{|x - f(x)|^2} \pm \frac{\sqrt{(f(x)(x - f(x))^2) - |x - f(x)|^2\,(|f(x)|^2 - 1)}}{|x - f(x)|^2}$$

Since $|x - f(x)|^2 \neq 0$,

$$\sqrt{(f(x)(x - f(x))^2) - |x - f(x)|^2(|f(x)|^2 - 1)} \geq |f(x)(x - f(x))| \geq 0,$$

ant hence t is a smooth function of x.

Thus g is a smooth map from B^n to ∂B^n. By Theorem 10.8.3, this is a contradiction.

Therefore there is $x \in B^n$ such that $f(x) = x$. □

10.8. Exercises

1. Show that the Brouwer fixed point theorem is false for the open ball $\overset{\circ}{B^n} = \{(x_1, \cdots, x_n) \in \mathbb{R}^n \mid x_1^2 + \cdots + x_n^2 < a\}$ where $a > 0$.

2. Find a map which the fixed point in the Brouwer theorem is a boundary point.

3. Prove the Brouwer theorem for maps of an interval $[a, b]$ where $a < b$.

Chapter 11

De Rham Cohomology

(1) Fundamental Theorem of Calculus :

$$\int_a^b d(f(x)) = \int_{\partial[a,b]} f(x) = f(b) - f(a)$$

(2) Green theorem :

$$\iint_\Omega d(f_1(x,y)dx + f_2(x,y)dy) = \int_{\partial\Omega} (f_1(x,y)dx + f_2(x,y)dy)$$

(3) Divergence Theorem :

$$\iiint_\Omega \text{div} F dv = \iint_{\partial\Omega} (F \cdot n) dA$$

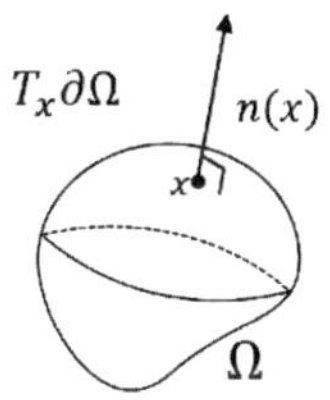

where $F = (f_1, f_2, f_3)$, n is the outward unit normal vector field on $\partial\Omega$.

(4) Stokes Theorem :

$$\begin{aligned}&\iint_{\Omega}(\text{curl}F\cdot n)dA\\&=\iint_{\Omega}(\frac{\partial f_2}{\partial x}-\frac{\partial f_1}{\partial y})dx\wedge dy+(\frac{\partial f_3}{\partial y}-\frac{\partial f_2}{\partial z})dy\wedge dz\\&\quad+(\frac{\partial f_1}{\partial z}-\frac{\partial f_3}{\partial x})dz\wedge dx\\&=\int_{\partial\Omega}f_1dx+f_2dy+f_3dz\end{aligned}$$

where $F := (f_1, f_2, f_3)$,

$\text{div}F := \frac{\partial f_1}{\partial x} + \frac{\partial f_2}{\partial y} + \frac{\partial f_3}{\partial z} = \nabla \cdot F$,

$\text{curl}F := (\frac{\partial f_3}{\partial y} - \frac{\partial f_2}{\partial z}, \frac{\partial f_1}{\partial z} - \frac{\partial f_3}{\partial x}, \frac{\partial f_2}{\partial x} - \frac{\partial f_1}{\partial y}) = \nabla \times F$.

11.1 Exterior Algebra

Let V be a k-dimensional real vector space, $V^p = \underbrace{V \times \cdots \times V}_{p-\text{times}}$.

Definition 11.1.1. A **p-tensor** on V is a multilinear

$$T : V^p \longrightarrow \mathbb{R}.$$

Example 11.1.2. (1) A **1-tensor** $f : V \longrightarrow \mathbb{R}$ is a linear map.

(2) The **inner product** $< \cdot, \cdot >: \mathbb{R}^k \times \mathbb{R}^k \longrightarrow \mathbb{R}$ is a 2-tensor on $\mathbb{R}^k$.

(3) The determinant $\det : V^k \longrightarrow \mathbb{R}$ is a k-tensor.

(4) If T and S are p-tensors, then λT and $T + S$ are also p-tensors. The set of all p-tensors on V, $\overset{p}{\otimes} V^*$ is a **vector space** where

$$V^* = \{f : V \longrightarrow \mathbb{R} \text{ is linear}\}$$

is the **dual vector space** of V.

Definition 11.1.3. Let $T \in \overset{p}{\otimes} V^*$ and $S \in \overset{q}{\otimes} V^*$. Define

$$(T \otimes S)(v_1, \cdots, v_p, r_{p+1}, \cdots, v_{p+q})$$
$$:= T(v_1, \cdots, v_p) S(v_{p+1}, \cdots, v_{p+q}).$$

Then $(T \otimes S) \in \overset{p+q}{\otimes} V^*$ is a $(p+q)$-tensor on V. The $T \otimes S$ is called the **tensor product** of T and S.

Remark 11.1.4. The tensor product $\otimes$ satisfies the associative and distributive laws, i.e.,

(1) $(T_1 \otimes T_2) \otimes T_3 = T_1 \otimes (T_2 \otimes T_3)$

(2) $T \otimes (S_1 + S_2) = T \otimes S_1 + T \otimes S_2$.

Theorem 11.1.5. *Let $\{e_1, \cdots, e_k\}$ be a basis of V^*. Then*

$$\{e_{i_1} \otimes \cdots \otimes e_{i_p} \mid 1 \leq i_1, \cdots, i_p \leq k\}$$

is a basis of $\overset{p}{\otimes} V^*$, $\dim \overset{p}{\otimes} V^* = k^p$.

Proof. For $1 \leq i_1, \cdots, i_p \leq k$, let $I = (i_1, \cdots, i_p)$, and $e_I = e_{i_1} \otimes \cdots \otimes e_{i_p}$. Let $\{v_1, \cdots, v_k\}$ be a basis of V, and $v_I = (v_{i_1}, \cdots, v_{i_p})$. Then

$$e_I(v_J) = \begin{cases} 1, & I = J, \\ 0, & I \neq J. \end{cases}$$

Two p-tensors $S = T$ iff for every $I = (i_1, \cdots, i_p)$, $S(v_I) = T(v_I)$.

Then $S = \sum_I S(v_I) e_I$, since for each I, $S(v_I) = (\sum_I S(v_I) e_I)(v_I)$. Thus $\overset{p}{\otimes} V^*$ is generated by $\{e_I\}_I$. If $\sum a_I e_I = 0$, then for each I

$$(\sum a_I e_I)(v_I) = a_I = 0.$$

So $\{e_I\}_I$ is linearly independent. Thus $\{e_I\}_I$ is a basis of $\overset{p}{\otimes} V^*$. $\square$

Definition 11.1.6. (1) A p-tensor T is **alternative**

$$T(v_1 \cdots, v_i, \cdots, v_j, \cdots, v_p) = -T(v_1, \cdots, v_j, \cdots, v_i, \cdots, v_p).$$

(2) $S_p = \{f : \{1, \cdots, p\} \longrightarrow \{1, \cdots, p\}$ is bijective $\}$ is the **permutation group of order** p.

Fact : Any permutation $\pi \in S_p$ can be written as a product of transpositions.

A permutation π is **even** (**odd**) if the number of transpositions is even (odd). Denote by

$$(-1)^\pi := \begin{cases} 1, & \text{if } \pi : \text{ even} \\ -1, & \text{if } \pi : \text{ odd.} \end{cases}$$

(3) For a p-tensor T, $\pi \in S_p$ define T^π by

$$T^\pi(v_1, \cdots, v_p) := T(v_{\pi(1)}, \cdots, v_{\pi(p)}).$$

Remark 11.1.7. (1) The determinant det is alternative.

(2) The inner product $< \, , \, >$ is not alternative.

(3) If T is an alternative p-tensor, then $T^\pi = (-1)^\pi T$.

Indeed, $T^\pi(v_1, \cdots, v_p) = T(v_{\pi(1)}, \cdots, v_{\pi(p)})$
$= (-1)^\pi T(v_1, \cdots, v_p)$.

If $\pi = (i, j)$, then

$$\begin{aligned} T(v_1, \cdots, v_j, \cdots, v_i, \cdots, v_p) &= T^\pi(v_1, \cdots, v_i, \cdots, v_j, \cdots, v_p) \\ &= (-1)^\pi T(v_1, \cdots, v_i, \cdots, v_j, \cdots, v_p) \\ &= -T(v_1, \cdots, v_i, \cdots, v_j, \cdots, v_p). \end{aligned}$$

(4) If T is a p-tensor, then

$$\mathrm{Alt}(T) := \frac{1}{p!} \sum_{\pi \in S_p} (-1)^\pi T^\pi$$

is an alternative p-tensor.

Proof. For any permutation $\sigma \in S_P$,

$$\begin{aligned}
\Big[\mathrm{Alt}(T)\Big]^{\sigma} &= \Big(\frac{1}{p!}\sum_{\pi\in S_p}(-1)^{\pi}T^{\pi}\Big)^{\sigma} \\
&= \frac{1}{p!}\sum_{\pi\in S_p}(-1)^{\pi}\Big(T^{\pi}\Big)^{\sigma} \\
&= \frac{1}{p!}\sum_{\pi\in S_p}(-1)^{\sigma}(-1)^{\pi}(-1)^{\sigma}\,T^{\pi\circ\sigma} \\
&= (-1)^{\sigma}\frac{1}{p!}\sum_{\imath\in S_p}(-1)^{\pi\circ\sigma}\,T^{\pi\circ\sigma} \\
&= (-1)^{\sigma}\mathrm{Alt}(T).
\end{aligned}$$

□

(5) If T is an alternative p-tensor, then

$$\begin{aligned}
\mathrm{Alt}(T) &= \frac{1}{p!}\sum_{\pi\in S_p}(-1)^{\pi}T^{\pi} \\
&= \frac{1}{p!}\sum_{\pi\in S_p}(-1)^{\pi}(-1)^{\pi}T = T.
\end{aligned}$$

(6) The set of all alternative p-tensor $\wedge^p(V^*)$ is closed under addition and scalar multiplication, i.e.,

$$\text{if } T_1,\, T_2 \in \wedge^p(V^*) \text{ and } \lambda\in\mathbb{R}, \text{ then } T_1+\lambda T_2 \in \wedge^p(V^*).$$

Thus $\wedge^p(V^*)$ is a vector space.

Definition 11.1.8. The **wedge product** $T\wedge S$ of $T \in \wedge^p(V^*)$ and $S\in\wedge^q(V^*)$ is defined by

$$T\wedge S := \mathrm{Alt}(T\otimes S) \in \wedge^{p+q}(V^*).$$

Since $\mathrm{Alt}\big((\lambda T_1+T_2)\otimes S\big) = \lambda\mathrm{Alt}\big(T_1\otimes S\big)+\mathrm{Alt}\big(T_2\otimes S\big)$,

$$\big(T_1+\lambda T_2\big)\wedge S = T_1\wedge S+\lambda\, T_2\wedge S.$$

Lemma 11.1.9. *If* $\operatorname{Alt}(T) = 0$, *then* $T \wedge S = 0 = S \wedge T$.

Proof. Let $T \in \wedge^p(V^*)$, $S \in \wedge^q(V^*)$. We may think $S_p \subset S_{p+q}$ $(1, \cdots, p, p+1, \cdots, p+q)$, by fixing $(p+1, \cdots, p+q)$.

$$\begin{aligned}\sum_{\pi \in S_p} (-1)^\pi (T \otimes S)^\pi &= \Big(\sum_{\pi \in S_p} (-1)^\pi T^\pi \Big) \otimes S \\ &= \operatorname{Alt}(T) \otimes S = 0.\end{aligned}$$

Since $S_{p+q} = \bigcup_{\sigma \in S_{p+q}} S_p \circ \sigma$,

$$\sum_{\pi \in S_p} (-1)^{\pi \circ \sigma} (T \otimes S)^{\pi \circ \sigma} = (-1)^\sigma \Big(\sum_{\pi \in S_p} (-1)^\pi (T \otimes S)^\pi \Big)^\sigma = 0.$$

Thus

$$\begin{aligned}T \wedge S &= \operatorname{Alt}(T \otimes S) \\ &= \frac{1}{(p+q)!} \sum_{\pi \in S_{p+q}} (-1)^\pi (T \otimes S)^\pi = 0.\end{aligned}$$

Similarly, we have $S \wedge T = 0$. □

Theorem 11.1.10. $(T \wedge S) \wedge R = T \wedge (S \wedge R)$.

Proof. Since $\operatorname{Alt}(T \wedge S - T \otimes S) = \operatorname{Alt}(T \wedge S) - \operatorname{Alt}(T \otimes S) = 0$,

$$\begin{aligned}&(T \wedge S) \wedge R - \operatorname{Alt}(T \otimes S \otimes R) \\ &\quad = \operatorname{Alt}\big((T \wedge S) \otimes R\big) - \operatorname{Alt}\big(T \otimes S \otimes R\big) \\ &\quad = \operatorname{Alt}\big((T \wedge S - (T \otimes S)) \otimes R\big) \\ &\quad = \operatorname{Alt}\Big[\big(T \wedge S - T \otimes S\big) \wedge R\Big] = 0.\end{aligned}$$

Similarly, $T \wedge (S \wedge R) = \operatorname{Alt}(T \otimes S \otimes R)$.
Thus $T \wedge (S \wedge R) = (T \wedge S) \wedge R$. □

By Theorem 11.1.10, $T \wedge S \wedge S = \mathrm{Alt}(T \otimes S \otimes R)$. Let

$$T = \sum t_{i_1 \cdots i_p} e_{i_1} \otimes \cdots \otimes e_{i_p}$$

be a p-tensor. If T is alternative, then

$$\begin{aligned} T &= \mathrm{Alt}(T) \\ &= \sum t_{i_1 \cdots i_p} \mathrm{Alt}(e_{i_1} \otimes \cdots \otimes e_{i_p}) \\ &= \sum t_{i_1 \cdots i_p} e_{i_1} \wedge \cdots \wedge e_{i_p}. \end{aligned}$$

Let $I = (i_1, \cdots, i_p)$ and $e_I = e_{i_1} \wedge \cdots \wedge e_{i_p}$.
Then $\{e_I\}_I$ generates $\wedge^p(V^*)$.

Remark 11.1.11. (1) $\{e_I\}_I$ is not linearly independent.
Since if ϕ, $\psi \in \wedge^1(V^*)$, then

$$\phi \wedge \psi = \frac{1}{2}(\phi \otimes \psi - \psi \otimes \phi) = -\psi \wedge \phi,$$

and $\phi \wedge \phi = 0$. Thus the wedge product $\wedge$ is anticommutative on $\wedge^1$.

(2) If I and J are the same set but has different order, then $e_I = \pm e_J$.

(3) If I has same elements, then $e_I = 0$.
Let $\wedge_p = \{I = (i_1, \cdots, i_p) \,|\, 1 \le i_1 < i_2 < \cdots, i_p \le k\}$. Then $\{e_I, \,|\, I \subset \wedge_p\}$ are linearly independent.

Indeed, if $\sum_{I \in \wedge_p} a_I e_I = 0$ and $\{v_I \,|\, I \in \wedge_p\}$ where $\{v_1, \cdots, v_k\}$ is a dual basis of $\{e_1, \cdots, e_k\}$ in V, then

$$0 = (\sum_{I \in \wedge_p} a_I e_I)(v_I) = \frac{1}{p!} a_I.$$

Thus $a_I = 0$ for any $I \in \wedge_p$.

Theorem 11.1.12. *If $\{e_1, \cdots, e_k\}$ is a basis of V^*, then*

$$\{e_I = e_{i_1} \wedge \cdots \wedge e_{i_p} \,|\, 1 \le i_1 < \cdots < i_p \le k\}$$

is a basis of $\wedge^p(V^*)$ and the dimension is

$$\dim \wedge^p(V^*) = {}_kC_p = \frac{k!}{p!(k-p)!}.$$

Corollary 11.1.13. *If $T \in \wedge^p(V^*)$ and $S \in \wedge^q(V^*)$, then*

$$T \wedge S = (-1)^{pq} S \wedge T.$$

Let V be a k-dimensional real vector space. Define $\wedge^0(V^*) = \mathbb{R}$, $\wedge^1(V^*) = V^*$, $\cdots$, $\wedge^k(V^*) =< e_1 \wedge \cdots \wedge e_k >\simeq \mathbb{R}$. The direct sum

$$\wedge(V^*) = \wedge^0(V^*) \oplus \cdots \oplus \wedge^k(V^*)$$

is an algebra under $+$ and $\wedge$ with identity $1 \in \wedge^0(V^*) = \mathbb{R}$.

Definition 11.1.14. $(\wedge(V^*), +, \wedge)$ is called the **exterior algebra** of V^*.

Let $A : V \longrightarrow W$ be a linear map. Then $A^* : W^* \longrightarrow V^*$ extends to exterior algebras by $T \in \wedge^p(W^*)$, $v_1, \cdots, v_p \in V$

$$(A^*T)(v_1, \cdots, v_p) := T(Av_1, \cdots, Av_p).$$

The A^* is a linear map and $A^*(T \wedge S) = A^*T \wedge A^*S$.
Thus $A^* : \wedge(W^*) \longrightarrow \wedge(V^*)$ is an algebra homomorphism.
If $B : W \longrightarrow U$ is a linear map, then

$$(BA)^* = A^*B^*.$$

Let $A : V \longrightarrow V$ be an isomorphism and $\dim_{\mathbb{R}} V = k$.
Then $A^* : \wedge^k(V^*) \longrightarrow \wedge^k(V^*)$ is an isomorphism of one-dimensional vector space.

For each nonzero $T \in \wedge^k(V^*)$ there is $\lambda \in \mathbb{R}$ such that $A^*(T) = \lambda T$. Find λ.

Solution: Let $B : V \longrightarrow \mathbb{R}^k$ be a linear isomorphism, and $B^*(\det) = T$. Then

$$A^*B^*(\det) = A^*(T) = \lambda T = \lambda B^*(\det),$$
$$B^{*^{-1}}A^*B^*(\det) = B^{*^{-1}}\lambda B^*(\det) = \lambda(\det).$$

Let $\{a_1, \cdots, a_k\}$ be the standard basis of $\mathbb{R}^k$. Then

$$(BAB^{-1})^*(\det)(a_1, \cdots, a_k) = \det(BAB^{-1}) = \det A,$$

$\lambda \det(a_1, \cdots, a_k) = \lambda$.
Thus $\det(A) = \lambda$.

Theorem 11.1.15. *Let $A : V \longrightarrow V$ be a linear isomorphism and $\dim V = k$.*

(1) *If $T \in \wedge^k(V^*)$ then $A^*(T) = \det(A) \cdot T$.*

(2) *If $\phi_1, \cdots, \phi_k \in \wedge^1(V^*)$, then*

$$\begin{aligned} A^*(\phi_1 \wedge \cdots \wedge \phi_k) &= A^*\phi_1 \wedge \cdots \wedge A^*\phi_k \\ &= \det(A) \cdot \phi_1 \wedge \cdots \wedge \phi_k. \end{aligned}$$

11.1. Exercises

1. Let V be a k-dimensional vector space. Prove :

(1) if $\phi_1, \cdots, \phi_p \in V^*$ are linear dependent, then $\phi_1 \wedge \cdots \wedge \phi_p = 0$.

(2) if $v_1, \cdots, v_p \in V$ are linear dependent, then

$$(\phi_1 \wedge \cdots \wedge \phi_p)(v_1, \cdots, v_p) = 0.$$

(3) if $\phi_1, \cdots, \phi_p \in V^*$ and $v_1, \cdots, v_p \in V$, then

$$(\phi_1 \wedge \cdots \wedge \phi_p)(v_1, \cdots, v_p) = \frac{1}{p!} \det\big(\phi_i(v_j)\big).$$

(4) Let $A : V \longrightarrow V$ be a linear map. Then the following are equivalent.

(i) $Av \cdot Aw = v \cdot w$ for all $v, w \in V$,

(ii) A sends orthonormal bases to orthonormal bases.

(iii) $A \cdot {}^tA = I_V$.

2. Write out $\mathrm{Alt}(\phi_1 \otimes \phi_2 \otimes \phi_3 \otimes \phi_4)$ for $\phi_1, \phi_2, \phi_3, \phi_4 \in V^*$.

3. Let $v \in V$. Prove that

$$\wedge^p V \xrightarrow{v\wedge} \wedge^{p+1} V \xrightarrow{v\wedge} \wedge^{p+2} V$$

is an exact sequence.

11.2 Differential Form

Let X be a smooth k-dimensional manifolds. For each point $x \in X$,
T_xX: the tangent space at x,
T_x^*X: the cotangent space at x,
$TX = \bigcup_{x\in X} T_xX$: the tangent bundle of X,
$T^*X = \bigcup_{x\in X} T_x^*X$: the cotangnet bundle of X.

$$\begin{array}{c} \wedge^p T^*X = \bigcup_{x\in X} \wedge^p T_x^*X \\ \Big\downarrow \pi \\ X \end{array}$$

where π is a projection map, i.e., $\pi(\wedge^p T_x^*X) = x$ for all $x \in X$.

Definition 11.2.1. A **p-form** ω on X is a section of the vector bundle

$$\pi : \wedge^p T^*X \longrightarrow X,$$

i.e., $\omega : X \longrightarrow \wedge^p T^*X$, $\omega(x) \in \wedge^p T_x^*X$ for each $x \in X$.
$\omega(x) : T_xX \times \cdots \times T_xX \longrightarrow \mathbb{R}$ is an alternative p-tensor.

Let $\Omega^p(X)$ be the set of all p-forms on X.

(1) If $\omega_1, \omega_2 \in \Omega^p(X)$, $\lambda \in \mathbb{R}$, then

$$(\omega_1 + \omega_2)(x) = \omega_1(x) + \omega_2(x),$$

$$(\lambda\omega_1)(x) = \lambda\omega_1(x) \quad \text{for each } x \in X,$$

and $\omega_1 + \omega_2$, $\lambda\omega_1 \in \Omega^p(X)$.

(2) if $\omega \in \Omega^p(X)$, $\theta \in \Omega^q(X)$ and

$$(\omega \wedge \theta) = \omega(x) \wedge \theta(x),$$

then $\omega \wedge \theta \in \Omega^{p+q}(X)$ and $\omega \wedge \theta = (-1)^{pq}\theta \wedge \omega$.

(3) $\Omega^0(X) := \{f : X \longrightarrow \mathbb{R} \text{ is smooth map}\}$ is the space of 0-forms on X. For $f \in \Omega^0(X)$, $df_x : T_xX \longrightarrow \mathbb{R}$ is a linear map. Thus $df_x \in T_x^*X$, and $df \in \Omega^1(X)$. The df is called the **differential** of f.

Example 11.2.2. Let $X = \mathbb{R}^k$. The coordinate functions x_i are given by $x_1, \cdots, x_k : \mathbb{R}^k \longrightarrow \mathbb{R}$. Their differentials $dx_1, \cdots, dx_k$ are 1-forms on $\mathbb{R}^k$. For each point $z \in \mathbb{R}^k$,

$$T_z\mathbb{R}^k = \mathbb{R}^k, \quad dx_j(z) \in T_z^*\mathbb{R}^k = (\mathbb{R}^k)^*, \quad dx_j(a_1, \cdots, a_k) = a_j.$$

Thus $dx_1(z), \cdots, dx_k(z)$ is the standard basis of $(\mathbb{R}^k)^*$.

For $1 \le i_1 < \cdots < i_p \le k$, $I = (i_1, \cdots, i_p)$, $dx_I = dx_{i_1} \wedge \cdots \wedge dx_{i_p}$ is a p-form on $\mathbb{R}^k$.

Theorem 11.2.3. *Let $U \subset \mathbb{R}^k$ be an open set, and $I = (i_1, \cdots, i_p)$, $1 \le i_1 < \cdots < i_p \le k$, $f_I : U \longrightarrow \mathbb{R}$ smooth. Then a p-form on U is $\omega = \sum_I f_I dx_I$.*

Let $\phi : \mathbb{R}^k \longrightarrow \mathbb{R}$ be a smooth function. Then its derivative $d\phi = \sum_{i=1}^k \frac{\partial \phi}{\partial x_i} dx_i$ is a 1-form on $\mathbb{R}^k$.

Let $f : X \longrightarrow Y$ be a smooth map and ω a p-form on Y, i.e. $\omega \in \Omega^p(Y)$. If $f(x) = y$, then $df_x : T_xX \longrightarrow T_yY$. For $\omega(y) \in \wedge^p T_y^* Y$ and $(df_x)^* : T_y^*Y \longrightarrow T_x^*X$,

$$(df_x)^* : \wedge^p T_y^* Y \longrightarrow \wedge^p T_x^* X$$

is defined by

$$\big((df_x)^*\omega(y)\big)(v_1, \cdots, v_p) = \omega(y)\big(df_x(v_1), \cdots, df_x(v_p)\big),$$

for every $v_1, \cdots, v_p \in T_xX$.

Definition 11.2.4. Let $f : X \longrightarrow Y$ be smooth. Then $f^* : \Omega^p(Y) \longrightarrow \Omega^p(X)$ defined by

$$f^*(\omega)(x) := (df_x)^*\omega(f(x)), \quad \forall \omega \in \Omega^p(Y), \quad \forall x \in X$$

is called the **pullback** of ω by f.
In particular, $f^* : \Omega^0(Y) \longrightarrow \Omega^0(X)$ is given by

$$f^*(\omega) = \omega \circ f : X \xrightarrow{\ f\ } Y \xrightarrow{\ \omega\ } \mathbb{R}.$$

Theorem 11.2.5. *Let maps* $X \xrightarrow{f} Y \xrightarrow{g} Z$ *be smooth. Then*

$$\Omega^p(Z) \xrightarrow{g^*} \Omega^p(Y) \xrightarrow{f^*} \Omega^p(X)$$

and

(1) $f^*(\omega_1 + \omega_2) = f^*\omega_1 + f^*\omega_2$,

(2) $f^*(\omega \wedge \theta) = f^*(\omega) \wedge f^*\theta$,

(3) $(g \circ f)^*\omega = f^*(g^*\omega)$. □

(1) Let $U \subset \mathbb{R}^k$, $V \subset \mathbb{R}^\ell$ be open, $f : U \longrightarrow V$ smooth, $(x_1, \cdots, x_k) \in U$, $(y_1, \cdots, y_\ell) \in V$ coordinate systems and $f = (f_1 \cdots, f_\ell)$. Its derivative

$$df_x = \Big(\frac{\partial f_i}{\partial x_j}(x)\Big)$$

is a $\ell \times k$-matrix. Then $(df_x)^*$ is its transpose, and

$$f^*(dy_i) = (df_x)^*(dy_i) = \sum_{j=1}^{k} \frac{\partial f_i}{\partial x_j} dx_j = df_i.$$

If $\omega = \sum_I a_I dy_I$, then

$$\begin{aligned} f^*\omega &= \sum_I f^*(a_I) df^*(dy_I) = \sum_I f^*(a_I) df_I \\ &= \sum_I a_I \circ f \, df_{i_1} \wedge \cdots \wedge df_{i_p}. \end{aligned}$$

Thus f^* is determined by 0-form and 1-forms df_i, $i = 1, \cdots, \ell$.

(2) Let $f : U \longrightarrow V$ be a diffeomorphism, $f(x) = y$ and $\omega = dy_1 \wedge \cdots \wedge dy_k$. Then

$$\begin{aligned} (f^*\omega)(x) &= df_x^*(dy_1 \wedge \cdots \wedge dy_k)(f(x)) \\ &= df_x^*(dy_1)(y) \wedge \cdots \wedge df_x^*(dy_k)(y) \\ &= df_{1x} \wedge \cdots \wedge df_{kx} \\ &= \det(df_x)\, dx_1(x) \wedge \cdots \wedge dx_k(x). \end{aligned}$$

Thus $df^*(dy_1 \wedge \cdots \wedge dy_k) = \det(df)\, dx_1 \wedge \cdots \wedge dx_k$.

(3) Let $U \subset \mathbb{R}^k$ be open. Then $\omega = \sum_I a_I dx_I$ is smooth on $U \Longleftrightarrow a_I$ is smooth on U for each I. If $f : U \longrightarrow V$ is smooth and ω a smooth form, then $f^*\omega$ is a smooth form.

Definition 11.2.6. A form ω on X is **smooth** if for each parametrization

$$\phi : U \subset \mathbb{R}^k \longrightarrow X,$$

the pullback $\phi^*\omega$ is smooth.

11.2. Exercises

1. Let $f : X \longrightarrow Y$ be a smooth map, $\Omega^*(X)$ the set of all differential forms on X. Then show that

(1) $f^* : \Omega^*(Y) \longrightarrow \Omega^*(X)$ is an algebra homomorphism.

(2) $d\big(f^*(\omega)\big) = f^*\big(d(\omega)\big)$, $\omega \in \Omega^*(Y)$

(3) $f^*(\omega)(v_1, \cdots, v_p)(x) = \omega_{f(x)}\big(df(v_1(x)), \cdots, df(v_p(x))\big)$ for $\omega \in \Omega^p(Y)$ and vector fields $v_1, \cdots, v_p$ on X.

11.3 Integration on Manifold

Recall the integration on Euclidean space.

Theorem 11.3.1. *Let $f : U \longrightarrow V$ be a diffeomorphism between open subsets in $\mathbb{R}^k$, and $g : V \longrightarrow \mathbb{R}$ an integrable function. Then*

$$\int_V g\, dy_1 \cdots dy_k = \int_U (g \circ f)\, |\det(df)|\, dx_1 \cdots dx_k.$$

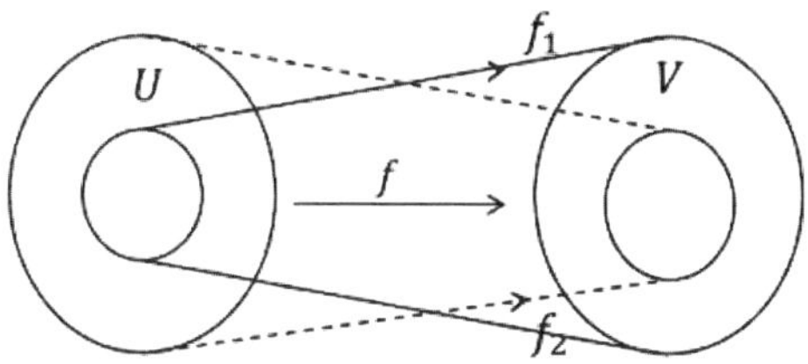

For A k-form $\omega = g dy_1 \wedge \cdots \wedge dy_k$ on an open subset $V \subset \mathbb{R}^k$, define

$$\int_V \omega := \int_V g\, dy_1 \cdots dy_k.$$

Since $f^*\omega = (g \circ f) \det(df)\, dx_1 \wedge \cdots \wedge dx_k$,

if f preserves the orientation on V, then $\det(df) > 0$,

if f reverses the orientation on V, then $\det(df) < 0$.

Definition 11.3.2. A k-form $\omega = g\, dy_1 \wedge \cdots \wedge dy_k$ is called **integrable** if g is integrable.

Theorem 11.3.3. *Let $f : U \longrightarrow V$ be a diffeomorphism between open subsets U and V in $\mathbb{R}^k$ or $\mathbb{H}^k$, ω an integrable k-form on V. Then*

$$\int_V \omega = \begin{cases} \int_U f^*\omega, & \text{if } f \text{ preserves orientation,} \\ -\int_U f^*\omega, & \text{if } f \text{ reverses orientation.} \end{cases}$$

Let X be a k-dimensional manifold with boundary, and ω a k-form on X.

Definition 11.3.4. (1) The **support** of $\omega := \overline{\{x \in X \,|\, \omega(x) \neq 0\}} := \mathrm{supp}(\omega)$.

(2) ω is **compactly supported** if $\mathrm{supp}(\omega)$ is compact.

At first, we assume that $h : U \subset \mathbb{H}^k \longrightarrow W \subset X$ is an orientation preserving parametrization such that $\mathrm{supp}(\omega) \subset W$, and $h^*\omega$ is a compactly supported k-form on U.

Definition 11.3.5. The **integral** of ω on X is defined by

$$\int_X \omega := \int_U h^*\omega.$$

Remark 11.3.6. If $g : V \longrightarrow W$ is another parametrization, then $f = h^{-1} \circ g : V \longrightarrow W \longrightarrow U$ is an orientation preserving diffeomorphism. By Theorem 11.3.3

$$\int_U h^*\omega = \int_V f^*h^*\omega = \int_V g^*\omega.$$

Thus $\int_X \omega$ is well-defined.

Let $\mathcal{W} = \{W_\alpha \,|\, h_\alpha : U_\alpha \longrightarrow W_\alpha \subset X \text{ is a parametrization}\}$ be an open cover of X. Let us assume that $\mathcal{W}$ is locally finite and $\{\psi_\alpha : X \longrightarrow \mathbb{R}\}$ be a partition of unity and subordinate to $\mathcal{W}$.

Definition 11.3.7. Let ω be a k-form with compact support. Then

$$\int_X \omega = \sum_\alpha \int_X \rho_\alpha \omega$$

is called the **integral** of ω over X.

Theorem 11.3.8. *The integral $\int_X \omega$ is independent to the partition of unity.*

Proof. By Remark 11.3.6 $\int_X \rho_\alpha\omega$ is well-defined.
Since $\sum_\alpha \rho_\alpha(x) = 1$ for each $x \in X$ and h_α^*, $\int_{\mathbb{R}^k}$ are linear,

$$\int_X \omega = \sum_\alpha \int_X \rho_\alpha\omega.$$

Suppose that $\{\rho'_\beta\}_\beta$ is another partition of unity.
For each α, $\int_X \rho_\alpha\omega = \sum_\beta \int_X \rho'_\beta\rho_\alpha\omega$, and for each β, $\int_X \rho'_\beta\omega = \sum_\alpha \int_X \rho_\alpha\rho'_\beta\omega$.
Thus

$$\begin{aligned}\sum_\alpha \int_X \rho_\alpha\omega &= \sum_\alpha\sum_\beta \int_X \rho'_\beta\rho_\alpha\omega \\ &= \sum_\beta\sum_\alpha \int_X \rho_\alpha\rho'_\beta\omega = \sum_\beta \int_X \rho'_\beta\omega.\end{aligned}$$

□

Remark 11.3.9. (1) $\int_X(\omega_1 + \omega_2) = \int_X \omega_1 + \int_X \omega_2$.

(2) $\int_X \lambda\omega = \lambda \int_X \omega$.

Theorem 11.3.10. *Let $f : X \longrightarrow Y$ be an orientation preserving diffeomorphism. Then*

$$\int_Y \omega = \int_X f^*\omega.$$

Proof. Let $\mathcal{V} = \{V_\alpha \,|\, \psi_\alpha : V'_\alpha \longrightarrow V_\alpha \subset Y$ is a parametrization$\}$, $f : U \subset X \longrightarrow Y$ be a locally finite covering of Y and $\{\rho_\alpha\}$ a partition of unity and subordinate to $\mathcal{V}$. Then

$$\mathcal{U} = \{U_\alpha \,|\, U_\alpha = f^{-1}(V_\alpha), \quad V_\alpha \in \mathcal{V}\}$$

is a locally finite open covering of X and $\{\rho_\alpha \circ f\}$ is a partition of unity and subordinate to $\mathcal{U}$. Thus we have

$$\begin{aligned}\int_Y \omega &= \sum_\alpha \int_Y \rho_\alpha\omega = \sum_\alpha \int_{V'_\alpha} \psi_\alpha^*(\rho_\alpha\omega) \\ &= \sum_\alpha \int_{V'_\alpha} (f \circ f^{-1} \circ \psi_\alpha)^*(\rho_\alpha\omega) = \sum_\alpha \int_{V'_\alpha} (f^{-1} \circ \psi_\alpha)^* f^*(\rho_\alpha\omega) \\ &= \sum_\alpha \int_{V'_\alpha} (f^{-1} \circ \psi_\alpha)^*(\rho_\alpha \circ f) f^*(\omega)\end{aligned}$$

$$= \sum_{\alpha} \int_X (\rho_\alpha \circ f) f^*(\omega)$$
$$= \int_X f^*\omega,$$

since $\{f^{-1} \circ \psi_\alpha : V_\alpha' \longrightarrow U_\alpha$ is a parametrization in $X\}$. □

Definition 11.3.11. Let $Z \subset X$ be a ℓ-dimensional submanifold and ω a ℓ-form on X. Suppose $\text{supp}(\omega) \cap Z$ is compact and $i : Z \longrightarrow X$ is the inclusion. Define the integral of ω on Z by

$$\int_Z \omega := \int_Z i^*\omega.$$

Example 11.3.12. (1) Let $\omega = f_1 dx_1 + f_2 dx_2 + f_3 dx_3$ be a 1-form in $\mathbb{R}^3$. Let a path $\gamma : I = [0,1] \longrightarrow X$ be a parametrization of $C = \gamma(I) \subset X$. Then $\int_C \omega = \int_I \gamma^*\omega$.

If $\gamma(t) = (\gamma_1(t), \gamma_2(t), \gamma_3(t))$, $\forall t \in I$, then $\gamma^* dx_i = d\gamma_i = \frac{d\gamma_i}{dt} dt$ and

$$\int_C \omega = \int_I \gamma^*\omega = \sum_{i=1}^{3} \int_0^1 f_i(\gamma(t)) \cdot \frac{d\gamma_i}{dt} dt = \int_0^1 F \cdot dr$$

is the **line integral** of a vector field $F = (f_1, f_2, f_3)$ on $\mathbb{R}^3$ along the curve C.

(2) Let $\omega = f_1 dx_2 \wedge dx_3 + f_2 dx_3 \wedge dx_1 + f_3 dx_1 \wedge dx_2$ be a 2-form with a compact support.

Let $g : \mathbb{R}^2 \longrightarrow \mathbb{R}$ be a smooth function. Then the graph is $S := \{(x_1, x_2, g(x_1, x_2)) \in \mathbb{R}^3 \,|\, (x_1, x_2) \in \mathbb{R}^2\}$, its parametrization $h : \mathbb{R}^2 \longrightarrow S$, is given by

$$h(x_1, x_2) = (x_1, x_2, g(x_1, x_2)).$$

Also $h^*(dx_1 \wedge dx_2) = dx_1 \wedge dx_2$,

$$h^*(dx_2 \wedge dx_3) = dx_2 \wedge dtg = dx_2 \wedge (\frac{\partial g}{\partial x_1} dx_1 + \frac{\partial g}{\partial x_2} dx_2)$$
$$- \frac{\partial g}{\partial x_1} dx_1 \wedge dx_2,$$

$h^*(dx_3 \wedge dx_1) = dg \wedge dx_1 = -\frac{\partial g}{\partial x_2} dx_1 \wedge dx_2$. This

$$\begin{aligned}\int_S \omega &= \int_{\mathbb{R}^2} h^*\omega \\ &= \int_{\mathbb{R}^2} \Big(-\frac{\partial g}{\partial x_1} h^*(f_1) - \frac{\partial g}{\partial x_3} h^*(f_2) + h^*(f_3) \Big) dx_1 \wedge dx_2 \\ &= \int_{\mathbb{R}^2} \Big(-\frac{\partial g}{\partial x_1}, -\frac{\partial g}{\partial x_2}, 1 \Big) h^*(f) dx_1 \wedge dx_2 \\ &= \int_{\mathbb{R}^2} \mathbf{n} \cdot h^*(f) dx_1 \wedge dx_2\end{aligned}$$

where

$$\mathbf{n} = (-\frac{\partial g}{\partial x_1}, -\frac{\partial g}{\partial x_2}, 1) = h_{x_1} \times h_{x_2} = \begin{vmatrix} \mathbf{i} & \mathbf{j} & \mathbf{k} \\ 1 & 0 & \frac{\partial g}{\partial x_1} \\ 0 & 1 & \frac{\partial g}{\partial x_2} \end{vmatrix}$$

$f = (f_1, f_2, f_3)$, $\mathbf{n}(x) \perp T_x S$, $x = (x_1, x_2, g(x_1, x_2))$.
Let $dA := |\mathbf{n}| dx_1 \wedge dx_2$ which is called the **area form**, $U := \frac{\mathbf{n}}{|\mathbf{n}|}$. Then

$$\int_S \omega = \int_{\mathbb{R}^2} U \cdot h^*(f) dA.$$

11.3. Exercises

1. Let $\gamma : [a, b] \longrightarrow X$ be a smooth curve and $f : X \longrightarrow \mathbb{R}$ a smooth map. Show that

$$\int_a^b \gamma^*(df) = f[\gamma(b)] - f[\gamma(a)].$$

2. Let $e : \mathbb{R} \longrightarrow S^1$ be $e(t) = (\cos t, \sin t)$ and ω a 1-form on S^1. Show that

$$\int_{S^1} \omega = \int_0^{2\pi} e^*\omega.$$

3. Let $\omega(x, y) = \frac{xdy}{x^2+y^2} - \frac{ydx}{x^2+y^2}$ be a 1-form on $\mathbb{R}^n - \{0\}$. Show that

(1) $d\omega = 0$.

(2) Calculate $\int_{S^1} \omega$.

(3) On $\{(x, y) \in \mathbb{R}^2 \mid x > 0\}$, $d(\tan^{-1} \frac{y}{x}) = \omega$.

(4) There is no smooth map $f : S^1 \longrightarrow \mathbb{R}$ such that $df = \omega$.

11.4 Exterior Derivative

Definition 11.4.1. Let $\omega = \sum a_I dx_I$ be a p-form on an open subset of Euclidean space $\mathbb{R}^k$. The $(p+1)$-form $d\omega = \sum da_I \wedge dx_I$ is called the **exterior derivative** of ω.

Theorem 11.4.2. *The exterior derivative d on the space of forms defined in open set of $\mathbb{R}^k$ or $\mathbb{H}^k$ is the unique operator satisfying the followings :*

(1) $d(\omega_1 + \omega_2) = d\omega_1 + d\omega_2$, $df = \sum_i \frac{\partial f}{\partial x_i} dx_i$,

(2) $d(\omega \wedge \theta) = d\omega \wedge \theta + (-1)^p \omega \wedge d\theta$, *where ω is a p-form,*

(3) $d(d\omega) = 0$.

Proof. (1) By definition it is obvious.
(2) If $\omega = \sum_I a_I dx_I$ and $\theta = \sum_J b_J dx_J$, then

$$\begin{aligned}\omega \wedge \theta &= (\sum_I a_I dx_I) \wedge (\sum_J b_J dx_J) \\ &= \sum_{I,J} (a_I b_J) dx_I \wedge dx_J.\end{aligned}$$

$$\begin{aligned}d(\omega \wedge \theta) &= \sum_{I,J} d(a_I b_J) \wedge dx_I \wedge dx_J \\ &= \sum_{I,J} (da_I b_J + a_I db_J) \wedge dx_I \wedge dx_J \\ &= \sum_{I,J} da_I b_J \wedge dx_I \wedge dx_J + \sum_{I,J} a_I db_J \wedge dx_I \wedge dx_J \\ &= (\sum_I da_I \wedge dx_I) \wedge \sum_J b_J dx_J \\ &\quad + (-1)^p (\sum_I a_I dx_I) \wedge \sum_J db_J \wedge dx_J \\ &= d\omega \wedge \theta + (-1)^p \omega \wedge d\theta.\end{aligned}$$

(3) For a p-form $\omega = \sum_I a_I dx_I$,

$$d\omega = \sum_I da_I \wedge dx_I = \sum_I (\sum_{i=1}^k \frac{\partial a_I}{\partial x_i} dx_i) \wedge dx_I.$$

$$\begin{aligned} d(d\omega) &= \sum_I \Big[\sum_i (\sum_j \frac{\partial^2 a_I}{\partial x_j \partial x_i} dx_j) \wedge dx_i\Big] \wedge x_I \\ &= \sum_I \Big(\sum_{i,j} \frac{\partial^2 a_I}{\partial x_j \partial x_i} dx_j \wedge dx_i\Big) \wedge dx_I \\ &= 0, \end{aligned}$$

since $\frac{\partial^2 a_I}{\partial x_j \partial x_i} = \frac{\partial^2 a_I}{\partial x_i \partial x_j}$ and $dx_j \wedge dx_i = -dx_j \wedge dx_j$.

Uniqueness of d satisfying (1) $\sim$ (3). Suppose d' is another operator on smooth forms such that (1) $\sim$ (3). Then $d'(dx_i) \overset{(1)}{=} d'(d'x_i) \overset{(3)}{=} 0$, and

$$\begin{aligned} d'(dx_I) &= d(dx_{i_1} \wedge \cdots \wedge dx_{i_p}) \\ &\overset{(2)}{=} d'(dx_{i_1}) \wedge dx_{i_2} \wedge \cdots \wedge dx_{i_p} + \cdots + dx_{i_1} \wedge (-1)^{p-1} d'(dx_{i_p}) \\ &= 0. \end{aligned}$$

For a p-form $\omega = \sum a_I dx_I$,

$$\begin{aligned} d'\omega &= \sum (d'a_I \wedge dx_I + a_I d'(dx_I)) \\ &= \sum d'a_I \wedge dx_I = \sum da_I \wedge dx_I = d\omega. \end{aligned}$$

□

Theorem 11.4.3. *Let U, V be open in $\mathbb{R}^k$ or $\mathbb{H}^k$ and $g : U \longrightarrow V$ be a diffeomorphism. Then*

$$d(g^*\omega) = g^*(d\omega).$$

Proof. Let $d' = g^{-1^*} \circ d \circ g^*$. Then d' satisfies (1) $\sim$ (3) of Theorem 11.4.2.

(1) Let $f : V \longrightarrow \mathbb{R}$, then

$$\begin{aligned} d'(f) &= g^{-1^*} d(g^* f) = g^{-1^*} d(f \circ g) \\ &= d(g^{-1^*}(f \circ g)) = d(f \circ g \circ g^{-1}) = df. \end{aligned}$$

(2), (3) are obvious by definition.

By uniqueness, $d = d' = g^{-1^*} \circ d \circ g^*$.
Thus $g^* \circ d = d \circ g^*$. □

Let $g : U \longrightarrow V$ be an orientation preserving diffeomorphism between open subsets of $\mathbb{R}^k$. Then

$$\int_V \omega = \int_U g^*\omega$$

and $d \circ g^* = g^* \circ d$.

Definition 11.4.4. Let $\omega \in \Omega^p(X)$ be a p-form on X, and $\phi : U \longrightarrow X$ a parametrization. Then we define the **exterior derivative** of ω on $\phi(U)$ by $d\omega := \phi^{-1^*} d(\phi^*\omega)$.

Let $\psi : V \longrightarrow X$ be another parametrization and $\phi(U) \cap \psi(V) \neq \emptyset$, and $g = \phi^{-1} \circ \psi$. Then $g^* d(\phi^*\omega) = d(g^*\phi^*\omega) = d(\psi^*\omega)$ and

$$\psi^{-1^*}\big(d(\psi^*\omega)\big) = \psi^{-1^*} g^* d(\phi^*\omega) = \phi^{-1^*} d(\phi^*\omega).$$

Thus $d\omega$ is well-defined on X.

Theorem 11.4.5. *Let d be the exterior derivative on the space of smooth forms on X. Then*

(1) $d(\omega_1 + \omega_2) = d\omega_1 + d\omega_2$.

(2) *If ω is a p-form on X, then*

$$d(\omega \wedge \theta) = d\omega \wedge \theta + (-1)^p \omega \wedge d\theta.$$

(3) $d(d\omega) = 0$.

(4) *If $f : X \longrightarrow \mathbb{R}$ is a smooth function, then on parametrization*

$$df = \sum_i \frac{\partial f}{\partial x_i} dx_i.$$

(5) *If* $g : Y \longrightarrow X$ *is a diffeomorphism, then* $d \circ g^* = g^* \circ d$.

Proof. On Euclidean space $\mathbb{R}^k$ we have proved Theorem 11.4.2. Since on manifold $d\omega = {\phi^{-1}}^* d(\phi^*\omega)$ we can check Theorem 11.4.5.

(5) $g^* d\omega = g^* {\phi^{-1}}^* d\phi^*\omega = (\phi^{-1}g)^* d(g^{-1}\phi)^* g^*\omega = dg^*\omega.$ □

Theorem 11.4.6. *Let* $g : Y \longrightarrow X$ *be a smooth map and* ω *a form on* X. *Then*

$$d(g^*\omega) = g^*(d\omega).$$

Proof. (1) If $\omega : X \longrightarrow \mathbb{R}$ is a 0-form, then by the chain rule, $dg^*\omega = g^* d\omega$.

(2) If $\omega = df$, $f : X \longrightarrow \mathbb{R}$, then

$$d\omega = d(df) = 0 \quad \text{and} \quad g^*(d\omega) = 0.$$

Also $d(g^*\omega) = d(g^* df) = d(dg^*(f)) = 0$.

Thus theorem holds for $\omega = df$.

(3) If ω and θ satisfy Theorem 11.4.6, then $\omega \wedge \theta$ satisfies Theorem 11.4.6.

Indeed, $dg^*(\omega \wedge \theta) = d(g^*\omega \wedge g^*\theta)$

$$\begin{aligned} &= dg^*\omega \wedge g^*\theta + (-1)^p g^*\omega \wedge dg^*\theta \\ &= g^* d\omega \wedge g^*\theta + (-1)^p g^*\omega \wedge g^* d\theta \\ &= g^* \left[d\omega \wedge \theta + (-1)^p \omega \wedge (d\theta) \right] \\ &= g^* d(\omega \wedge \theta). \end{aligned}$$

(4) Locally $\omega = \sum a_I dx_I$, the ω satisfies Theorem 11.4.6. Thus globally the theorem holds.

□

Example 11.4.7 (On $\mathbb{R}^3$). (1) A 0-form $f : \mathbb{R}^3 \longrightarrow \mathbb{R}$ is a smooth function. Then

$$df = \frac{\partial f}{\partial x_1} dx_1 + \frac{\partial f}{\partial x_2} dx_2 + \frac{\partial f}{\partial x_3} dx_3$$

is a 1-form.

$$(\frac{\partial f}{\partial x_1}, \frac{\partial f}{\partial x_2}, \frac{\partial f}{\partial x_3}) := \text{grad}(f) = \nabla(f),$$

where $\nabla = (\frac{\partial}{\partial x_1}, \frac{\partial}{\partial x_2}, \frac{\partial}{\partial x_3})$ is called the **gradient vector field** of f.

(2) For a 1-form $\omega = f_1 dx_1 + f_2 dx_2 + f_3 x_3$, the derivative

$$\begin{aligned} d\omega &= df_1 \wedge dx_1 + df_2 \wedge dx_2 + df_3 \wedge dx_3 \\ &= (\frac{\partial f_3}{\partial x_2} - \frac{\partial f_2}{\partial x_3}) dx_2 \wedge dx_3 \\ &\quad + (\frac{\partial f_1}{\partial x_3} - \frac{\partial f_3}{\partial x_1}) dx_3 \wedge dx_1 + (\frac{\partial f_2}{\partial x_1} - \frac{\partial f_1}{\partial x_2}) dx_1 \wedge dx_2. \end{aligned}$$

$$\begin{aligned} \left(\frac{\partial f_3}{\partial x_2} - \frac{\partial f_2}{\partial x_3}, \frac{\partial f_1}{\partial x_3} - \frac{\partial f_3}{\partial x_1}, \frac{\partial f_2}{\partial x_1} - \frac{\partial f_1}{\partial x_2}\right) \\ = (\frac{\partial}{\partial_{x_1}}, \frac{\partial}{\partial_{x_2}}, \frac{\partial}{\partial_{x_3}}) \times (f_1, f_2, f_3) \\ = \nabla \times F, \quad F = (f_1, f_2, f_3) \\ = \text{curl}(F). \end{aligned}$$

Here curl(F) is called the **curl** of $F = (f_1, f_2, f_3)$.

(3) For a 2-form $\omega = f_1 dx_2 \wedge dx_3 + f_2 dx_3 \wedge dx_1 + f_3 dx_1 \wedge dx_2$,

$$\begin{aligned} d\omega &= \left(\frac{\partial f_1}{\partial x_1} + \frac{\partial f_2}{\partial x_2} + \frac{\partial f_3}{\partial x_3}\right) dx_1 \wedge dx_2 \wedge dx_3 \\ &= (\frac{\partial}{\partial x_1}, \frac{\partial}{\partial x_2}, \frac{\partial}{\partial x_3}) \cdot (f_1, f_2, f_3) dx_1 \wedge dx_2 \wedge dx_3 \\ &= \nabla \cdot F\, dx_1 \wedge dx_2 \wedge dx_3, \quad F = (f_1, f_2, f_3) \\ &= \text{div}(F)\, dx_1 \wedge dx_2 \wedge dx_3. \end{aligned}$$

Here $\text{div}(f_1, f_2, f_3) = \frac{\partial f_1}{\partial x_1} + \frac{\partial f_2}{\partial x_2} + \frac{\partial f_3}{\partial x_3}$ is called the **divergence** of F.

(4) 3-form $\omega = f dx_1 \wedge dx_2 \wedge dx_3$, then $d\omega = 0$.

11.4. Exercises

1. Calculate the exterior derivatives of the following forms on $\mathbb{R}^3$.

(1) On S^2,

$$\frac{-ydx}{x^2+y^2+z^2+1}+\frac{ydz}{x^2+y^2+z^2+1}+\frac{-xdz}{x^2+y^2+z^2+1}.$$

(2) If $f, g : \mathbb{R}^3 \longrightarrow \mathbb{R}$ are smooth, then compute $f\,dg$.

(3) $e^{xyz}dx \wedge dy + (\sin y^2 + xe^z)dy \wedge dz$.

(4) $f(x\,dy \wedge dz + xyz\,dz \wedge dx)$ where $f : \mathbb{R}^3 \longrightarrow \mathbb{R}$ is a smooth map.

2. On $\mathbb{R}^3$,

(1) If $f : \mathbb{R}^3 \longrightarrow \mathbb{R}$ is a smooth map, show $\operatorname{curl}\big(\operatorname{grad}(f)\big) = 0$.

(2) If F is a vector field on $\mathbb{R}^3$, then show

$$\operatorname{div}\big(\operatorname{curl}(F)\big) = 0.$$

11.5 De Rham Cohomology

Let X be a k-dimensional differentiable manifold. Then the space of p-forms of X is

$$\Omega^p(X) := \Gamma(\wedge^p T^*M) = \{\omega : X \longrightarrow \wedge^p T^*X \text{ is a section}\}$$

i.e., $\omega \in \Omega^p(X)$, $\omega(x) \in \wedge^p T_x^* M$ for each $x \in X$,
$\omega(x) : T_xX \times \cdots \times T_xX \longrightarrow \mathbb{R}$ is an alternative p-tensor.

$$\begin{aligned}
\omega(x) &= \sum_I a_I dx_I, \\
& dx_{i_1} \wedge \cdots \wedge dx_{i_p}(v_1, \cdots, v_p) \\
&\quad = \frac{1}{p!} \sum_{\sigma \in s_p} (-1)^\sigma (dx_{i_{\sigma(1)}} \otimes \cdots \otimes dx_{i_{\sigma(p)}})(v_1, \cdots, v_p) \\
&= \frac{1}{p!} \sum_{\sigma \in s_p} (-1)^\sigma dx_{i_{\sigma(1)}}(v_1) \cdots \otimes dx_{i_{\sigma(p)}}(v_p) \\
&= \frac{1}{p!} \det(dx_{i_{\sigma(k)}} v_{(k)}).
\end{aligned}$$

The de Rham cochain complex of X is

$$0 \longrightarrow \Omega^0(X) \xrightarrow{d} \Omega^1(X) \xrightarrow{d} \cdots \xrightarrow{d} \Omega^p(X)$$

$$\xrightarrow{d} \cdots \xrightarrow{d} \Omega^k(X) \xrightarrow{d} 0,$$

where $\Omega^p(X)$ is an ∞-dimensional vector space, d is linear and $d^2 = 0$.

Definition 11.5.1. (1) $\omega \in \Omega^p(X)$ is **closed** if $d\omega = 0$.

(2) $\omega \in \Omega^p(X)$ is **exact** if $d\theta = \omega$ for some $\theta \in \Omega^{p-1}(X)$.

Remark 11.5.2. (1) Every exact form is closed.

(2) $\mathrm{curl}(\mathrm{grad}(f)) = 0$, $\quad \mathrm{div}(\mathrm{curl}F) = 0$.

Recall in $\mathbb{R}^3$:

$$0 \longrightarrow \Omega^0(\mathbb{R}^3) \xrightarrow{d} \Omega^1(\mathbb{R}^3) \xrightarrow{d} \Omega^2(\mathbb{R}^3) \xrightarrow{d} \Omega^3(\mathbb{R}^3) \xrightarrow{d} 0.$$

(i) $f \in \Omega^0(\mathbb{R}^3)$, $f : \mathbb{R}^3 \longrightarrow \mathbb{R}$, $(x_1, x_2, x_3) \mapsto f(x_1, x_2, x_3)$.

$$df = \frac{\partial f}{\partial x_1}dx_1 + \frac{\partial f}{\partial x_2}dx_2 + \frac{\partial f}{\partial x_3}dx_3 \in \Omega^1(\mathbb{R}^3)$$

$$\mathrm{grad}(f) := \nabla f = (\frac{\partial f}{\partial x_1}, \frac{\partial f}{\partial x_2}, \frac{\partial f}{\partial x_3}) \quad \text{is a vector field.}$$

$$\begin{aligned}
d(df) &= d(\frac{\partial f}{\partial x_1}) \wedge dx_1 + d(\frac{\partial f}{\partial x_2}) \wedge dx_2 + d(\frac{\partial f}{\partial x_3}) \wedge dx_3 \\
&= \Big[\frac{\partial^2 f}{\partial x_1 \partial x_1}dx_1 + \frac{\partial^2 f}{\partial x_2 \partial x_1}dx_2 + \frac{\partial^2 f}{\partial x_3 \partial x_1}dx_3\Big] \wedge dx_1 + \cdots \\
&= \Big[\frac{\partial^2 f}{\partial x_2 \partial x_3} - \frac{\partial^2 f}{\partial x_3 \partial x_2}\Big]dx_2 \wedge dx_3 + \Big[\frac{\partial^2 f}{\partial x_3 \partial x_1} - \frac{\partial^2 f}{\partial x_1 \partial x_3}\Big]dx_3 \wedge dx_1 \\
&\qquad + \Big[\frac{\partial^2 f}{\partial x_1 \partial x_2} - \frac{\partial^2 f}{\partial x_2 \partial x_1}\Big]dx_1 \wedge dx_2 \\
&= 0
\end{aligned}$$

(ii) Let $\omega = f_1 dx_1 + f_2 dx_2 + f_3 dx_3 \in \Omega^1(\mathbb{R}^3)$ be a 1-from. Let $F = (f_1, f_2, f_3)$ be a vector field on $\mathbb{R}^3$.

$$\begin{aligned}
d\omega &= \Big[\frac{\partial f_1}{\partial x_1}dx_1 + \frac{\partial f_1}{\partial x_2}dx_2 + \frac{\partial f_1}{\partial x_3}dx_3\Big] \wedge dx_1 + \cdots \\
&= \Big[\frac{\partial f_3}{\partial x_2} - \frac{\partial f_2}{\partial x_3}\Big]dx_2 \wedge dx_3 + \Big[\frac{\partial f_1}{\partial x_3} - \frac{\partial f_3}{\partial x_1}\Big]dx_3 \wedge dx_1 \\
&\qquad + \Big[\frac{\partial f_2}{\partial x_1} - \frac{\partial f_1}{\partial x_2}\Big]dx_1 \wedge dx_2 \in \Omega^2(\mathbb{R}^3)
\end{aligned}$$

$$\begin{aligned}
\mathrm{curl}(F) &:= \Big(\frac{\partial f_3}{\partial x_2} - \frac{\partial f_2}{\partial x_3}, \frac{\partial f_1}{\partial x_3} - \frac{\partial f_3}{\partial x_1}, \frac{\partial f_2}{\partial x_1} - \frac{\partial f_1}{\partial x_2}\Big) = \nabla \times F \\
&= \begin{vmatrix} \mathbf{i} & \mathbf{j} & \mathbf{k} \\ \frac{\partial}{\partial x_1} & \frac{\partial}{\partial x_2} & \frac{\partial}{\partial x_3} \\ f_1 & f_2 & f_3 \end{vmatrix}
\end{aligned}$$

is a vector field. Thus $d^2(f) = 0 \iff \mathrm{curl}(\mathrm{grad}(f)) = 0$.

(iii) For a 2-form $\sigma = f_1 dx_2 \wedge dx_3 + f_2 dx_3 \wedge dx_1 + f_3 dx_1 \wedge dx_2 \in \Omega^2(\mathbb{R}^3)$,

$$d\sigma = \Big(\frac{\partial f_1}{\partial x_1} + \frac{\partial f_2}{\partial x_2} + \frac{\partial f_3}{\partial x_3}\Big)dx_1 \wedge dx_2 \wedge dx_3 \in \Omega^3(\mathbb{R}^3).$$

Let $F = (f_1, f_2, f_3)$ be a vector field on $\mathbb{R}^3$.

$$\operatorname{div}(F) := \frac{\partial f_1}{\partial x_1} + \frac{\partial f_2}{\partial x_2} + \frac{\partial f_3}{\partial x_3} = \nabla \cdot F$$

is a function $\mathbb{R}^3$. For a vector field $F = (f_1, f_2, f_3)$,

$$\begin{aligned}
&\operatorname{div}(\operatorname{curl}(F)) = \operatorname{div}\Big(\frac{\partial f_3}{\partial x_2} - \frac{\partial f_2}{\partial x_3}, \frac{\partial f_1}{\partial x_3} - \frac{\partial f_3}{\partial x_1}, \frac{\partial f_2}{\partial x_1} - \frac{\partial f_1}{\partial x_2}\Big) \\
&= \Big[\frac{\partial^2 f_3}{\partial x_1 \partial x_2} - \frac{\partial^2 f_2}{\partial x_1 \partial x_3} + \frac{\partial^2 f_1}{\partial x_2 \partial x_3} - \frac{\partial^2 f_3}{\partial x_2 \partial x_1} + \frac{\partial^2 f_2}{\partial x_3 \partial x_1} - \frac{\partial^2 f_1}{\partial x_3 \partial x_2}\Big] \\
&= 0 = d\big[d(f_1 dx_1 + f_2 dx_2 + f_3 dx_3)\big].
\end{aligned}$$

(3) For a 1-form $\omega = \frac{-ydx}{x^2+y^2} + \frac{xdy}{x^2+y^2}$ on $\mathbb{R}^2 - \{0\}$,

$$\begin{aligned}
d\omega &= d\Big(\frac{-y}{x^2+y^2}\Big) \wedge dx + d\Big(\frac{x}{x^2+y^2}\Big) \wedge dy \\
&= \frac{-dy(x^2+y^2) + y2ydy}{(x^2+y^2)^2} \wedge dx + \frac{dx(x^2+y^2) - x2xdx}{(x^2+y^2)^2} \wedge dy \\
&= \frac{(-x^2 - y^2 + 2y^2)}{(x^2+y^2)^2} dy \wedge dx + \frac{x^2 + y^2 - 2x^2}{(x^2+y^2)^2} dx \wedge dy \\
&= \frac{x^2 + y^2 - 2y^2 + x^2 + y^2 - 2x^2}{(x^2+y^2)^2} dx \wedge dy = 0.
\end{aligned}$$

In polar coordinate of (x, y) in $\mathbb{R}^2 - \{0\}$, $\tan\theta = \frac{y}{x}$.

Then $\theta = \tan^{-1}\frac{y}{x}$ and

$$\begin{aligned}
d\theta &= d\Big(\tan^{-1}\frac{y}{x}\Big) \\
&= \frac{d(\frac{y}{x})}{1 + (\frac{y}{x})^2} \\
&= \frac{\frac{xdy - ydx}{x^2}}{\frac{x^2+y^2}{x^2}} \\
&= \frac{-ydx + xdy}{x^2+y^2} = \omega.
\end{aligned}$$

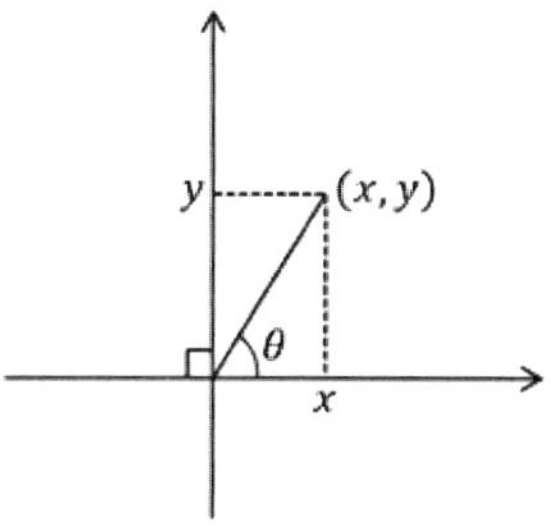

Definition 11.5.3.

$$H^p(X, \mathbb{R}) = \frac{\operatorname{Ker} d : \Omega^p(X) \longrightarrow \Omega^{p+1}(X)}{\operatorname{Im} d : \Omega^{p-1}(X) \longrightarrow \Omega^p(X)}$$

is called the **p-th de Rham cohomology** of X.

Let $f : X \longrightarrow Y$ be a smooth map. Since $f^* \circ d = d \circ f^*$, the diagram

$$\begin{array}{ccccccccc}
0 \longrightarrow \Omega^0(Y) & \xrightarrow{d} \cdots \xrightarrow{d} & \Omega^p(Y) & \xrightarrow{d} & \Omega^p(X) & \xrightarrow{d} & \Omega^{p+1}(X) & \longrightarrow \cdots \\
\downarrow f^* & & \downarrow f^* & & \downarrow f^* & & \downarrow f^* & \\
0 \longrightarrow \Omega^0(X) & \xrightarrow{d} \cdots \xrightarrow{d} & \Omega^{p-1}(X) & \xrightarrow{d} & \Omega^p(X) & \xrightarrow{d} & \Omega^{p+1}(X) & \longrightarrow \cdots
\end{array}$$

is commutative. So

$$f^*(\operatorname{Ker} d) \subset \operatorname{Ker}(d) \quad \text{and} \quad f^*(\operatorname{Im} d) \subset \operatorname{Im}(d).$$

Thus we have a well-defined linear map

$$f^\sharp : H^p(Y, \mathbb{R}) \longrightarrow H^p(X, \mathbb{R})$$

which is defined $f^\sharp([\omega]) = [f^*(\omega)]$.

Theorem 11.5.4. (1) *For* $I_X : X \longrightarrow X$, $I_X^\sharp = I_{H^p(X,\mathbb{R})}$, *for all* p.

(2) *For* $X \xrightarrow{f} Y \xrightarrow{g} Z$, $(g \circ f)^\sharp = f^\sharp \circ g^\sharp$.
Thus for each p,

$$H^p : \{\textit{manifolds}\} \longrightarrow \{\textit{vector spaces}\}$$

is a contravariant functor.

Indeed: $H^q(Z) \xrightarrow{g^\sharp} H^q(Y) \xrightarrow{f^\sharp} H^q(X)$,

$$\begin{aligned}(g \circ f)^\sharp([\omega]) &= [(g \circ f)^*(\omega)] = [f^*(g^*(\omega))] \\ &= f^\sharp[g^*(\omega)] = (f^\sharp \circ g^\sharp)([\omega]), \quad \forall [\omega] \in H^q(Z).\end{aligned}$$

Theorem 11.5.5. (1) *If* $\dim X < p$, *then* $H^p(X) = 0$.

(2) *If* X *has* n *connected components, then* $H^0(X) = \mathbb{R}^n$.

Proof. (1) If $\dim X < p$, then $\Omega^p(X) = \{0\}$ and $H^p(X) = \{0\}$.

(2) $\Omega^0(X) = \{f : X \longrightarrow \mathbb{R} \text{ is smooth}\}$

$df = 0 \iff f$ is constant on each component of X.

If X is connected, then

$$\operatorname{Ker} d = \{f \in \Omega^0(X) \mid f : X \longrightarrow \mathbb{R} \text{ is constant}\} \simeq \mathbb{R}.$$

If $X = \cup_{i=1}^n X_i$ is the disjoint union of connected components of X, then

$$H^0(X) = \oplus_{i=1}^n H^0(X_i) = \oplus_{i=1}^n \mathbb{R} = \mathbb{R}^n.$$

□

Let U be an open subset of $\mathbb{R}^k$, and a p-form $\omega \in \Omega^p(\mathbb{R} \times U)$. Then

$$\omega = \sum_I f_I(t,x)dt \wedge dx_I + \sum_J g_J(t,x)dx_J,$$

where $t \in \mathbb{R}$, $x = (x_1, \cdots, x_k) \in \mathbb{R}^k$, $I = (1 \leq i_1 < \cdots < i_{p-1} \leq k)$, $J = (1 \leq j_1 < \cdots < j_p \leq k)$.

Define an operator

$$P : \Omega^p(\mathbb{R} \times U) \longrightarrow \Omega^{p-1}(\mathbb{R} \times U),$$

by $P\omega(t,x) = \sum_I [\int_0^t f_I(s,x)ds]dx_I$.
Let $\phi : V \longrightarrow U$ be a diffeomorphism between open sets U and V in $\mathbb{R}^k$. Then $\Phi = \mathrm{id} \times \phi \ : \ \mathbb{R} \times V \longrightarrow \mathbb{R} \times U$ is a diffeomorphism.

Theorem 11.5.6. *$\Phi^* P\omega = P\Phi^*\omega$, for $\omega \in \Omega^p(\mathbb{R} \times U)$.*

Proof. Consider the following diagram :

$$\begin{array}{ccc} \Omega^p(\mathbb{R} \times U) & \xrightarrow{P} & \Omega^{p-1}(\mathbb{R} \times U) \\ \downarrow{\Phi^*} & & \downarrow{\Phi^*} \\ \Omega^p(\mathbb{R} \times V) & \xrightarrow{P} & \Omega^{p-1}(\mathbb{R} \times V). \end{array}$$

For $\omega = \sum_I f_I(t,x)dt \wedge dx_I + \sum_J g_J(t,x)x_J \in \Omega^p(\mathbb{R}\times U)$, and

$$\Phi = \mathrm{id} \times \phi = \mathrm{id} \times (\phi_1, \cdots, \phi_k).$$

$$\Phi^*(dt) = dt, \quad \Phi^*(dx_i) = d\phi_i.$$

Hence

$$\begin{aligned}\Phi^* P\omega &= \Phi^* \sum_I \Big(\int_0^t f_I(s,x)ds\Big)dx_I \\ &= \sum_I \Big(\int_0^t f_I(s,\phi(x))ds\Big)d\phi_I,\end{aligned}$$

and

$$\begin{aligned}P\Phi^*\omega &= P\Big(\sum_I f_I(t,\phi(x))dt \wedge d\phi_I + \sum_J g_J(t,\phi(x))d\phi_J\Big) \\ &= \sum_I \Big[\int_0^t f_I(s,\phi(x))ds\Big]d\phi_I.\end{aligned}$$

Thus $\Phi^*P = P\Phi^*$. □

Theorem 11.5.7. *For every manifold X there is a unique operator $P : \Omega^p(\mathbb{R}\times X) \longrightarrow \Omega^{p-1}(\mathbb{R}\times X)$ such that*

(1) *If $\phi : X \longrightarrow Y$ is a diffeomorphism and $\Phi = \mathrm{id}\times\phi : \mathbb{R}\times X \longrightarrow \mathbb{R}\times X$, then $\Phi^* \circ P = P \circ \Phi^*$.*

(2) *If U is open in $\mathbb{R}^k$ and a p-form $\omega \in \Omega^p(\mathbb{R}\times U)$*

$$\omega = \sum_I f_I(t,x)dt \wedge dx_I + \sum_J g_J(t,x)dx_J,$$

then

$$P\omega(t,x) = \sum_I \Big(\int_0^t f_I(s,x)ds\Big)dx_I.$$

Proof. Let $\psi : \mathbb{R}\times U \longrightarrow \mathbb{R}\times X$ be a parametrization. Then for $\omega \in \Omega^p(\mathbb{R}\times X)$, $\Psi^*(\omega) \in \Omega^p(\mathbb{R}\times U)$. By (2), $P\Psi^*(\omega) \in \Omega^{p-1}(\mathbb{R}\times U)$.

On $\mathrm{Im}\Psi \subset \mathbb{R}\times X$, we define $P(\omega) = \psi^{-1*}P\psi^*(\omega)$.

$$
\begin{array}{ccccc}
\Omega^p(\mathbb{R}\times Y) & & \xrightarrow{\quad P \quad} & & \Omega^{p-1}(\mathbb{R}\times Y) \\
\big\downarrow{\scriptstyle \Phi^*} & & & & \big\downarrow{\scriptstyle \Phi^*} \\
\Omega^p(\mathbb{R}\times X) & & \xrightarrow{\quad P \quad} & & \Omega^{p-1}(\mathbb{R}\times X) \\
 & \searrow{\scriptstyle \psi_1^*} & & \nearrow{\scriptstyle \psi_1^{-1*}} & \\
\big\downarrow{\scriptstyle \psi^*} & \Omega^p(\mathbb{R}\times V) & \xrightarrow{P} & \Omega^{p-1}(\mathbb{R}\times V) & \big\uparrow{\scriptstyle \psi^{-1*}} \\
\Omega^p(\mathbb{R}\times U) & & \xrightarrow{\quad P \quad} & & \Omega^{p-1}(\mathbb{R}\times U)
\end{array}
$$

where $\psi_1 : \mathbb{R}\times V \longrightarrow \mathbb{R}\times X$ is another parametrization Then

$$P\omega = \psi_1^{-1^*} P\psi_1^*(\omega).$$

On the common part of parametrizations :

$$
\begin{aligned}
\psi^{-1^*} P\psi^*\omega &= \psi_1^{-1^*}\psi_1^*\psi^{-1^*} P\psi^*\psi_1^{-1^*}\psi_1^*\omega \\
&= \psi_1^{-1^*}(\psi^{-1}\circ\psi_1)^* P(\psi_1^{-1}\circ\psi)^*\psi_1^*\omega \\
&\overset{\text{Thm 11.5.6}}{=} \psi_1^{-1^*}(\psi^{-1}\circ\psi_1)^*(\psi_1^{-1}\circ\psi)^* P\psi_1^*\omega \\
&= \psi_1^{-1^*} P\psi_1^*\omega.
\end{aligned}
$$

Thus $P : \Omega^p(\mathbb{R}\times X) \longrightarrow \Omega^{p-1}(\mathbb{R}\times X)$ is well-defined.

For $\omega \in \Omega^p(\mathbb{R}\times Y)$,

$$
\begin{aligned}
\Phi^* P(\omega) &= \Phi^*\psi^{-1^*} P\psi^*(\omega) \\
&= (\psi^{-1}\Phi)^* P\psi^*(\omega) \\
&= P(\psi^{-1}\Phi)^*\psi^*(\omega) \\
&= P\Phi^*(\omega).
\end{aligned}
$$

Thus $\Phi^* P = P\Phi^*$.

Since P is unique on Euclidean spaces $\mathbb{R}^k$, P is also unique on manifolds X. $\square$

Theorem 11.5.8. *Let $\pi : \mathbb{R}\times X \longrightarrow X$ be projection, and for each $a\in\mathbb{R}$, $i_a : X \longrightarrow \mathbb{R}\times X$, $i_a(x) = (a, x)$. Then*

$$dP\omega + Pd\omega = \omega - \pi^* i_a^*\omega.$$

Proof. The operators d and P are defined on parametrizations and are well-defined globally. Thus we may assume that X is an Euclidean space.

Let $\omega = \sum_I f_I(t,x)dt \wedge dx_I + \sum_J g_J(x,t)dx_J \in \Omega^p(\mathbb{R} \times X)$. Then $\pi^* i_a^* = \sum_J g_J(a,x)dx_J$.

$$\begin{aligned}
dP\omega + Pd\omega &= d\Big(\sum_I \big[\int_0^t f_I(s,t)ds\big]dx_I\Big) \\
&\quad + P\Big[\sum_I df_I(t,x) \wedge dt \wedge dx_I + \sum_J dg_J(t,x) \wedge dx_J\Big] \\
&= \sum_I f_I(t,x)dt \wedge dx_I + \sum_I \Big(\sum_i \frac{\partial}{\partial x_i} \int_a^t f_I(s,x)dx_i\Big) \wedge dx_I \\
&\quad + P\Big[\sum_I \sum_i \frac{\partial f_I(t,x)}{\partial x_i} dx_i \wedge dt \wedge dx_I + \sum_J \frac{\partial g_J(t,x)}{\partial t} dt \wedge dx_J\Big] \\
&= \sum_I f_I(t,x)dt \wedge dx_I + \sum_J g_J(t,x)dx_J - \sum g_J(a,x)dx_J \\
&= \omega - \pi^* i_a^* \omega.
\end{aligned}$$

□

Theorem 11.5.9 (Poincaré Lemma). $i_a^\sharp : H^p(\mathbb{R} \times X) \longrightarrow H^p(X)$ *and* $\pi^\sharp : H^p(X) \longrightarrow H^p(\mathbb{R} \times X)$ *are inverse each other.*

Proof. Consider maps :

$$X \xrightarrow{\ i_a\ } \mathbb{R} \times X \xrightarrow{\ \pi\ } X \xrightarrow{\ i_a\ } \mathbb{R} \times X.$$

Since $\pi \circ i_a = I_X$, by Theorem 11.5.4 (2) and (1),

$$i_a^\sharp \circ \pi^\sharp = (\pi \circ i_a)^\sharp = I_X^\sharp = I_{H^p(X)}.$$

For each $\omega \in H^p(\mathbb{R} \times X)$, by Theorem 11.5.8,

$$\omega - \pi^\sharp i_a^\sharp \omega = dP\omega + Pd\omega = dP\omega = 0$$

in $H^p(\mathbb{R} \times X)$. Thus $\pi^\sharp \circ i_a^\sharp = I_{H^p(\mathbb{R}\times X)}$. □

Theorem 11.5.10. (1) $H^p(\mathbb{R}^k) = 0$ *for* $p > 0$, $k \geq 0$.

(2) $H^0(\mathbb{R}^k) = \mathbb{R}$ *for* $k \geq 0$.

Proof. For all $k \geq 0$, by Theorem 11.5.9.

$$H^p(\mathbb{R}^k) \simeq H^p(\mathbb{R}^{k-1}) \simeq \cdots \simeq H^p(\mathbb{R}^0) \simeq H^p(\mathrm{pt}) = \begin{cases} 0, & p > 0 \\ \mathbb{R}, & p = 0. \end{cases}$$

□

Definition 11.5.11. Let $I = [0, 1]$.

(1) Two maps f, $g : X \longrightarrow Y$ are **homotopic** if there exists a smooth (homotopy) map $H : I \times X \longrightarrow Y$ such that $H(0, x) = f(x)$ and $H(1, x) = g(x)$. Denote it by $f \simeq g$. Then $\simeq$ is an equivalence relation on manifolds.

(2) Two spaces X and Y are **homotopic** (homotopically equivalent or the same homotopy type) if $f : X \longrightarrow Y$ and $g : Y \longrightarrow X$ such that $g \circ f \simeq I_X$ and $f \circ g \simeq I_Y$.

Theorem 11.5.12. *If* f, $g : X \longrightarrow Y$ *are homotopic, then* $f^\sharp = g^\sharp$.

Proof. Let $H : I \times X \longrightarrow Y$ be a homotopy such that $H(0, x) = f(x)$, $H(1, x) = g(x)$. Then $f = H \circ i_0$, $g = H \circ i_1$. So $f^\sharp = i_0^\sharp \circ H^\sharp$, $g^\sharp = i_1^\sharp \circ H^\sharp$ by Theorem 11.5.4.

By Theorem 11.5.9, $i_0^\sharp$ and $i_1^\sharp$ are inverse of π^* and $i_0^\sharp = i_1^\sharp$. Thus

$$f^\sharp = i_0^\sharp \circ H^\sharp = i_1^\sharp \circ H^\sharp = g^\sharp.$$

□

Theorem 11.5.13. *If X and a point $\{x\}$ are homotopic, then*

$$H^p(X) = \begin{cases} 0, & p > 0 \\ \mathbb{R}, & p = 0 \end{cases}$$

Proof. For any maps

$$\{x\} \xrightarrow{f} X \xrightarrow{g} \{x\},$$

$f \circ g \simeq I_X$, $g \circ f \simeq I_{\{x\}}$. By Theorem 11.5.12,

$$g^\sharp \circ f^\sharp = I_{H^p(X)}, \quad f^\sharp \circ g^\sharp = I_{H^p(\{x\})}.$$

Thus

$$H^p(X) \simeq H^p(\{x\}) = \begin{cases} 0, & p > 0 \\ \mathbb{R}, & p = 0 \end{cases}$$

□

Theorem 11.5.14.

$$H^p(S^k) = \begin{cases} \mathbb{R}^2, & p = k = 0 \\ \mathbb{R}, & k > 0 \text{ and } (p = k \text{ or } p = 0) \\ 0, & \text{otherwise.} \end{cases}$$

Proof. Prove Theorem 11.5.14 by induction on k.

(1) $k = 0$, $S^0 = \{-1, 1\} \Longrightarrow H^p(S^0) = \begin{cases} \mathbb{R}^2, & \text{if } p = 0, \\ 0, & \text{if } p \neq 0. \end{cases}$

(2) $k = 1$, S^1. $H^p(S^1) = \begin{cases} \mathbb{R}, & p = 0, 1 \\ 0, & p > 1. \end{cases}$

Show $H^1(S^1) = \mathbb{R}$.

$$\begin{array}{ccccccc} 0 & \longrightarrow & \Omega^0(S^1) & \xrightarrow{d} & \Omega^1(S^1) & \xrightarrow{d} & \Omega^2(S^1) = 0 \\ & & \| & & \| & & \\ & & \{f : S^1 \longrightarrow \mathbb{R}\} & & \operatorname{Ker} d & & \end{array}$$

Let θ be the polar coordinate $(0, 2\pi) \xrightarrow{\theta} S^1$ and $\frac{\partial}{\partial\theta}$ the unit vector field on S^1, $d\theta$ its dual 1-form.

Since θ is not globally defined, $d\theta$ is not exact.

$$\Omega^1(S^1) = \{g(\theta)d\theta \mid g : S^1 \longrightarrow \mathbb{R} \text{ is smooth}\}.$$

For any 1-form $g(\theta)d\theta \in \Omega^1(S^1)$, let $c = \frac{1}{2\pi}\int_0^{2\pi} g(\theta)d\theta$. Then

$$\begin{aligned} g(\theta)d\theta - cd\theta &= (g(\theta) - c)d\theta \\ &= \frac{d}{d\theta}\Big[\int_0^\theta g(s)ds - c\theta\Big]d\theta \\ &= d\Big[\int_0^\theta g(s)ds - c\Big]. \end{aligned}$$

Thus $[g(\theta)ds] = [c\,d\theta]$ in $H^1(S^1)$, $c = \frac{1}{2\pi}\int_0^{2\pi} g(\theta)d\theta$. Therefore,

$$H^1(S^1) = \frac{\Omega^1(S^1)}{\operatorname{Im} d} = \{c\,d\theta \mid c \in \mathbb{R}\} \simeq \mathbb{R}.$$

Hence $H^p(S^1) = \begin{cases} \mathbb{R}, & p = 0, 1 \\ 0, & \text{otherwise.} \end{cases}$

□

By induction we assume that Theorem 11.5.14 holds for S^{k-1}. Let $U_1 = S^k - \{N : \text{north pole}\}$, $U_2 = S^k - \{S : \text{south pole}\}$. Then $U_1 \simeq U_2 \simeq \mathbb{R}^k$ are diffeomorphic and $U_1 \cap U_2 \simeq \mathbb{R} \times S^{k-1}$.

Lemma 11.5.15. *For $p > 1$, $H^p(U_1 \cup U_2) \simeq H^{p-1}(U_1 \cap U_2)$ is an isomorphism.*

Proof. Let $\omega \in H^p(U_1 \cup U_2) = H^p(S^k)$. Then
on U_1, $\omega = d\phi_1$ for some $(p-1)$-form ϕ_1 on U_1, and
on U_2, $\omega = d\phi_2$ for some $(p-1)$-form ϕ_2 on U_2.
On $U_1 \cap U_2$, let $\nu = \phi_1 - \phi_2$. Then $d\nu = d\phi_1 - d\phi_2 = 0$. Thus

$$\nu \in H^{p-1}(U_1 \cap U_2).$$

Conversely, let $\{\rho_1, \rho_2\}$ be a partition of unity and subordinate to $\{U_1, U_2\}$ on S^k.

Let $\nu \in H^{p-1}(U_1 \cap U_2)$. Then $\phi_1 := \rho_1\nu$ is a $(p-1)$-form on U_1, and $\phi_2 := -\rho_2\nu$ is a $(p-1)$-form on U_2. On $U_1 \cap U_2$,

$$\phi_1 - \phi_2 = \rho_1\nu + \rho_2\nu = (\rho_1 + \rho_2)\nu = \nu.$$

Define a p-form ω on $U_1 \cup U_2 = S^k$ by

$$\omega := \begin{cases} d\phi_1 \text{ on } U_1, \\ d\phi_2 \text{ on } U_2. \end{cases}$$

Then on $U_1 \cap U_2$,

$$d\phi_1 - d\phi_2 = d(\phi_1 - \phi_2) = d\nu = 0.$$

Thus ω is well-defined on $U_1 \cup U_2$, and $d\omega = 0$.
Therefore $\omega \in H^p(U_1 \cup U_2)$. □

Continue the proof of Theorem 11.5.14.
For $p > 1$,

$$\begin{aligned} H^p(S^k) = H^p(U_1 \cup U_2) &\overset{\text{Theorem 11.5.14}}{=} H^{p-1}(U_1 \cap U_2) \\ &= H^{p-1}(\mathbb{R} \times S^{k-1}) = H^{p-1}(S^{k-1}). \end{aligned}$$

For $p = 1$, $H^1(S^1) = \mathbb{R}$ and Mayer-Vietoris sequence of

$$\begin{array}{ccc} S^2 = U_1 \cup U_2 & \longleftarrow & U_1 \simeq \mathbb{R}^2 \\ \uparrow & & \uparrow \\ \mathbb{R}^2 \simeq U_2 & \longleftarrow & U_1 \cap U_2 \simeq S^1, \end{array}$$

$$\begin{aligned} 0 &\longrightarrow H^0(S^2) \xrightarrow{\Gamma} H^0(U_1) \oplus H^0(U_2) \xrightarrow{\Phi} H^0(S^1) \xrightarrow{\delta} H^1(S^2) \\ &\longrightarrow H^1(U_1) \oplus H^1(U_2) \longrightarrow H^1(S^1) \xrightarrow{\simeq} H^2(S^2) \\ &\longrightarrow H^2(U_1) \oplus H^2(U_2) \longrightarrow \cdots \end{aligned}$$

is exact.

Since Γ is $1-1$ and $\operatorname{Im}\Gamma = \operatorname{Ker}\Phi = \mathbb{R}$, $\operatorname{Im}\Phi = \mathbb{R}$ and Φ is onto. Since $\operatorname{Im}\Phi = \operatorname{Ker}\delta = \mathbb{R}$ and $\delta = 0$, so $H^1(S^2) = 0$. Also $H^1(S^1) = H^2(S^2) = \mathbb{R}$.

Thus by induction for $k > 1$ and the relation in Lemma 11.5.15,

$$H^p(S^k) \simeq H^{p-1}(S^{k-1}) = \begin{cases} \mathbb{R}, & \text{if } p = 0, k \\ 0, & \text{otherwise.} \end{cases}$$

11.6 Stokes Theorem

Let X be a compact differentiable k-dimensional manifold. Then the cochain complex of X is

$$0 \longrightarrow \Omega^0(X) \xrightarrow{d} \Omega^1(X) \xrightarrow{d} \cdots \xrightarrow{d} \Omega^{k-1}(X) \xrightarrow{d} \Omega^k(X) \longrightarrow 0.$$

The chain complex of X is

$$0 \longrightarrow S_k(X) \xrightarrow{\partial} S_{k-1}(X) \xrightarrow{\partial} \cdots \xrightarrow{\partial} S_1(X) \xrightarrow{\partial} S_0(X) \longrightarrow 0.$$

Theorem 11.6.1 (Stokes Theorem). *Let X be a compact, differentiable, oriented k-dimensional manifold. If ω is a $(k-1)$-form of X, then*

$$\int_{\partial X} \omega = \int_X d\omega.$$

Proof. Since the integral $\int$ is linear we assume that the ω has compact support in a parametrization $\phi : U \longrightarrow X$.

(1) If $U \subset \mathbb{R}^k$ and $\phi(U) \cap \partial X = \emptyset$, then $\int_{\partial X} \omega = 0$, and

$$\int_X d\omega = \int_U \phi^*(d\omega) = \int_U (\phi^*\omega).$$

The $\phi^*\omega$ is a $(k-1)$-form on $\mathbb{R}^k$ and write

$$\phi^*\omega = \sum_{i=1}^{k} (-1)^{i-1} f_i dx_1 \wedge \cdots \wedge \widehat{dx_i} \wedge \cdots \wedge dx_k.$$

Then $d\phi^*\omega = \sum_{i=1}^k \frac{\partial f_i}{\partial x_i} dx_1 \wedge \cdots \wedge dx_k$.

By Fubini Theorem,

$$\begin{aligned}\int_{\mathbb{R}^k} d\phi^*\omega &= \sum_i \int_{\mathbb{R}^k} \frac{\partial f_i}{\partial x_i} dx_1 \wedge dx_2 \wedge \cdots \wedge dx_k \\ &= \sum_i \int_{\mathbb{R}^{k-1}} \Big(\int_{-\infty}^{\infty} \frac{\partial f_i}{\partial x_i} dx_i \Big) dx_1 \wedge \cdots \wedge \widehat{dx_i} \wedge \cdots \wedge dx_k.\end{aligned}$$

Let $g(x_i) = f(a_1, \cdots, x_i, \cdots, a_k)$ be a function of x_i. Then

$$\frac{\partial f_i}{\partial x_i} dx_i = \frac{dg(x_i)}{dx_i} dx_i = dg(x_i).$$

Taking integral we have

$$\int_{-\infty}^{\infty} \frac{\partial f_i}{\partial x_i} dx_i = \int_{-\infty}^{\infty} \frac{\partial f_i}{\partial x_i}(a_1, \cdots, x_i, \cdots, a_k) dx_i$$
$$= \int_{-\infty}^{\infty} \frac{dg}{dx_i} dx_i = \int_{-\infty}^{\infty} dg = 0,$$

since g has a compact support. Thus

$$\int_{\mathbb{R}^k} d\phi^* \omega = \int_X d\omega = 0.$$

(2) If $U \subset \mathbb{H}^k$, $\phi(U) \cap \partial X \neq \emptyset$, then

$$\int_X d\omega = \int_{\mathbb{H}^k} d\phi^* \omega = \int_{\mathbb{R}^{k-1}} \Big(\int_0^{\infty} \frac{\partial f_k}{\partial x_k} dx_k \Big) dx_1 \cdots dx_{k-1}$$
$$= \int_{\mathbb{R}^{k-1}} [-f_k(x_1, \cdots, x_{k-1}, 0)] dx_1 \cdots dx_{k-1},$$

since ω has a compact support, $\int_{-\infty}^{\infty} \frac{\partial f_i}{\partial x_i} = 0$, $i = 1, \cdots, k-1$. While

$$\int_{\partial X} \omega = \int_{\partial \mathbb{H}^k} \phi^* \omega = \int_{\partial \mathbb{H}^k} (-1)^{k-1} f_k(x_1, \cdots, x_k, 0) dx_1 \wedge \cdots \wedge dx_{k-1},$$

where $x_k = 0$ on $\partial \mathbb{H}^k$ and $dx_k = 0$.

Thus $\phi^* \omega = (-1)^{k-1} f_k(x_1, \cdots, x_{k-1}, 0) dx_1 \wedge \cdots \wedge dx_{k-1}$.

Let $\{e_1, \cdots, e_k\}$ be the standard ordered basis of $\mathbb{R}^k$. Then

$$\{-e_k, \text{standard basis of } \partial \mathbb{H}^k\} = \{e_1, \cdots, e_k\}.$$

Thus the orientation of $\partial \mathbb{H}^k = (-1)^k \{e_1, \cdots, e_k\}$.

$$\int_{\partial X} \omega = \int_{\mathbb{R}^{k-1}} (-1)^k (-1)^{k-1} f_k(x_1, \cdots, x_{k-1}, 0)\, dx_1 \cdots dx_{k-1}$$
$$= \int_{\mathbb{R}^{k-1}} \Big[- f_k(x_1, \cdots, x_{k-1}, 0) \Big] dx_1 \cdots dx_{k-1}.$$

Therefore $\int_{\partial X} \omega = \int_X d\omega$.

□

11.6. Exercises

1. Let $f : [a, b] \longrightarrow \mathbb{R}$ be a smooth map. Show

$$\int_a^d df = f(b) - f(a).$$

2. Let Ω be a compact domain in $\mathbb{R}^2$ with smooth boundary and f, $g : \Omega \longrightarrow \mathbb{R}$ smooth. Prove

$$\int_{\partial\Omega} f dx + g dy = \int_{\Omega} \Big(\frac{\partial g}{\partial x} - \frac{\partial f}{\partial y} \Big) dx dy.$$

3. Let Σ be a compact oriented manifold with boundary in $\mathbb{R}^3$, $\mathbf{n}$ the outward normal to Σ, and $V = (v_1, v_2, v_3)$ a smooth vector field around Σ. Prove

$$\int_{\Sigma} (\mathrm{curl} F \cdot \mathbf{n}) dA = \int_{\partial\Sigma} v_1 dx + v_2 dy + v_3 dz.$$

4. Let Ω be a compact domain in $\mathbb{R}^3$ with smooth boundary, and $V = (v_1, v_2, v_3)$ a smooth vector field on Ω. Show

$$\int_{\Omega} (\mathrm{div} V) dx dy dz = \int_{\partial\Omega} \mathbf{n} \cdot F dA,$$

where $\mathbf{n}$ is the outward normal vector field to $\partial\Sigma$.

5. Let $X = \partial M$, M a $(n+1)$-dimensional compact manifold, $F : M \longrightarrow Y$ smooth, and ω a closed n-form on Y. Show $\int_X F^*(\omega) = 0$.

6. Let g_0, $g_1 : X \longrightarrow Y$ be homotopic on a compact n-dimensional manifold X without boundary. Prove that if ω is a closed n-form on Y, then

$$\int_X g_0^* \omega = \int_X g_1^* \omega.$$

11.7 Degree Theorem

Theorem 11.7.1 (Degree Formula). *Let $f : X \longrightarrow Y$ be a smooth map between oriented compact connected k-dimensional manifolds. If ω is a k-form on Y, then*

$$\int_X f^*\omega = \deg(f) \cdot \int_Y \omega.$$

Theorem 11.7.2. *Let $\Omega \subset \mathbb{C}$ be a domain with smooth boundary $\partial\Omega$,*

$$p(z) = z^m + a_1 z^{m-1} + \cdots + a_m$$

a complex polynomial with no zero on $\partial\Omega$. Then the number of zeros (roots) in Ω with multiplicity is

$$\frac{1}{2\pi}\int_{\partial\Omega} d\,\mathrm{arg} p(z).$$

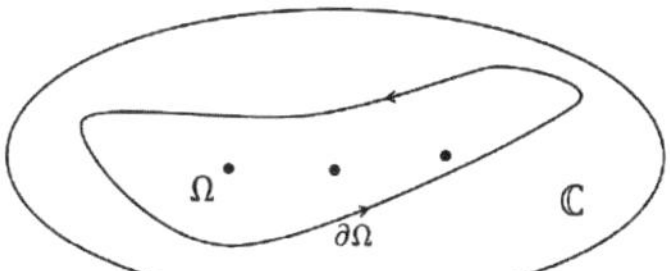

Proof. The argument of a complex number $\omega = re^{i\theta}$ is $\arg(\omega) = \theta$. In fact,

$$\omega = re^{i\theta} = re^{i(\theta+2n\pi)}, \quad n \in \mathbb{Z}, \quad \arg(\omega) = \theta + 2n\pi,$$

but the exterior derivative $d\arg(\omega) = d\theta$.

Since $\arg(\omega)$ is not well-defined on $\mathbb{C}$, the 1-form $d\theta$ is not exact. The 1-form

$$d\arg(p(z)) = dp^*\arg(z) = p^*d\arg(z)$$

is defined on $\mathbb{C}$ except $p(z) = 0$.

Let $f : \partial\Omega \longrightarrow S^1$ be defined by

$$f(z) = \frac{p(z)}{|p(z)|} = e^{i\mathrm{arg}(p(z))}.$$

Then $\deg(f) := \sharp$ of roots of $p(z)$ in Ω is called the **degree** of f.

Show

$$\int_{\partial\Omega} d\arg(p(z)) = 2\pi \cdot \deg(f).$$

Let $\omega = f(z) = e^{i\arg(p(z))}$. Then $\arg(\omega) = \arg(p(z))$, and

$$\begin{aligned}\arg(\omega) &= \arg(f(z)) = f^*\arg(z),\\ d\arg(p(z)) &= d\Big[f^*\arg(z)\Big]\\ &= f^*d\arg(z).\end{aligned}$$

Thus we have

$$\begin{aligned}\int_{\partial\Omega} d\arg(p(z)) &= \int_{\partial\Omega} f^*d\arg(z)\\ &\overset{Theorem\ 11.7.1}{=} \deg(f)\cdot\int_{S^1} d\arg(z).\end{aligned}$$

Consider a parametrization of S^1:

$$(0, 2\pi) \longrightarrow S^1 - \{1\}, \quad \theta \mapsto e^{i\theta}.$$

Then $d\arg\omega = d\theta$, and

$$\int_{S^1} d\arg(\omega) = \int_0^{2\pi} d\theta = 2\pi.$$

Therefore

$$\frac{1}{2\pi}\int_{\partial W} d\arg(p(z)) = \deg(f) = \sharp \text{ of roots of } \ p(z) \text{ in } \Omega.$$

□

Theorem 11.7.3. *Let $X = \partial W$, $f : X \longrightarrow Y$ have a smooth extension $F : W \longrightarrow Y$, X, Y, Z: compact, and* $\dim X = \dim Y = k$. *Then $\int_X f^*\omega = 0$, where ω is a k-form on Y.*

Proof.

$$\begin{aligned}\int_X f^*\omega &= \int_{\partial W} f^*\omega\\ &= \int_W d(f^*\omega) = \int_W f^*d\omega = 0,\end{aligned}$$

since $d\omega = 0$ and $\dim Y = k$. □

Corollary 11.7.4. *Let X, Y be compact oriented k-dimensional manifolds. Then for any k-form ω on Y*

$$\int_X f_0^*\omega = \int_X f_1^*\omega,$$

if f_0, $f_1 : X \longrightarrow Y$ are homotopic.

Proof. Let $F : I \times X \longrightarrow Y$ be a homotopy such that $F(0,x) = f_0(x)$, $F(1,x) = f_1(x)$. Since $\partial(I \times X) = \{1\} \times X - \{0\} \times X := X_1 - X_0$,

$$F|_{X_1} = f_1, \quad \text{and} \quad F|_{X_0} = f_0.$$

By Theorem 11.7.3

$$0 = \int_{\partial(I\times X)} (\partial F)^*\omega = \int_{X_1} f_1^*\omega - \int_{X_0} f_0^*\omega.$$

Thus $\int_X f_1^*\omega = \int_{X_1} f_1^*\omega = \int_{X_0} f_0^* = \int_X f_0^*\omega$. □

Lemma 11.7.5. *Let X, Y be compact oriented k-dimensional manifolds. If y is a regular value of $f : X \longrightarrow Y$, then there is a nbd U of y such that*

$$\int_X f^*\omega = \deg(f) \int_Y \omega$$

where ω is a k-form and $\mathrm{supp}(\omega) \subset U$.

Proof. Since y is a regular value of $f : X \longrightarrow Y$, there is a finite set $f^{-1}(y) = \{x_1, \cdots, x_n\}$ and a nbd U of y, a nbd V_i of x_i, $i = 1, 2, \cdots, n$ such that $f : V_i \longrightarrow U$ is a diffeomorphism, $i = 1, \cdots, n$.

If ω is a k-form and $\mathrm{supp}(\omega) \subset U$, then $f^*\omega$ has support in $f^{-1}(U) = \cup_{i=1}^n V_i$. Thus

$$\int_X f^*\omega = \sum_{i=1}^n \int_{V_i} f^*\omega.$$

Since $f : V_i \longrightarrow U$ is a diffeomorphism $i = 1, \cdots, n$,

$$\int_{V_i} f^*\omega = \varepsilon_i \int_U \omega,$$

where $\varepsilon_i = \begin{cases} 1, & \text{if } f : V_i \longrightarrow U \iff \text{preserves orientation} \\ -1, & \text{if } f : V_i \longrightarrow U \Longleftarrow \text{reverses orientation.} \end{cases}$

Thus $\deg(f) = \sum_{i=1}^{n} \varepsilon_i$, and

$$\int_X f^*\omega = \deg(f) \cdot \int_Y \omega.$$

□

Proof of Theorem 11.7.1. Let $y \in Y$ be a regular value of $f : X \longrightarrow Y$, and U a nbd of y as in Lemma 11.7.5.

Fact: For each point $z \in Y$ there exists a diffeomorphism $h_z : Y \longrightarrow Y$ such that $h_z(y) = z$ and h_z is isotopic to the identity I_Y,
i.e., there is $H : I \times Y \longrightarrow Y$ is a homotopy $H(0, \cdot) = h_z$, $H(1, \cdot) = I_Y$ and $H(t, \cdot)$ is a diffeomorphism for each $t \in I$.

There is an open cover $\{h_z(U) \,|\, z \in Y\}$ of Y. Since Y is compact, there is finite diffeomorphisms $h_1, \cdots, h_n$ such that $Y = \cup_{i=1}^{n} h_i(U)$. Let $\{\rho_i : Y \longrightarrow \mathbb{R}\}$ be a partition of unity and subordinate to $\{h_i(U)\}$. Then

$$\omega = \sum_{i=1}^{n} \rho_i\omega \quad \text{and} \quad \mathrm{supp}(\rho_i\omega) \subset h_i(U).$$

Since both sides of the degree formula

$$\int_X f^*\omega = \deg(f) \int_Y \omega$$

are linear in ω, it suffices to prove the formula for forms supported in some $h(U)$.

Assume that ω is a form supported in $h(U)$. Since $h \simeq \mathrm{id}_Y$, $h \circ f \simeq f$. By Corollary 11.7.4,

$$\int_X f^*\omega = \int_X (h \circ f)^*\omega = \int_X f^*(h^*\omega) \overset{Lemma\ 11.7.5}{=} \deg(f) \int_Y h^*\omega.$$

Since $h \simeq \mathrm{id}_Y$ and $\deg(h) = +1$,

$$\int_Y h^*\omega = \int_Y \omega.$$

Thus we have $\int_X f^*\omega = \deg(f) \int_Y \omega$. □

Appendix

We introduce the fact in the proof of Theorem 11.7.1.

Definition 11.7.6. (1) An **isotopy** is a homotopy in which each map h_t is a diffeomorphism.

(2) Two diffeomorphisms are **isotopic** if they can be joined by an isotopy.

(3) An isotopy is **compact supported** if the maps h_t are all equal to the identity outside some fixed compact set.

Isotopy Lemma. *For any given two points $y,\ z \in Y$ in a connected manifold Y, there exists a diffeomorphism $H : Y \longrightarrow Y$ such that $h(y) = z$ and h is isotopic to I_Y. Moreover, the isotopy may be taken to be compactly supported.*

Proof. (Hint) First construct h_t on $\mathbb{R}^1$.

Let $\rho : \mathbb{R} \longrightarrow \mathbb{R}$ be a smooth function such that $\begin{cases} \rho = 0 \text{ on } (-\varepsilon, \varepsilon)^c, \\ \rho(0) = 1 \end{cases}$

Define $h_t(x) = x + t\rho(x)z$. Then $h_t(x) = x$, if $x \notin (-\varepsilon, \varepsilon)$ or $t = 0$. Moreover it follows that $h_1(0) = z$.

In $\mathbb{R}^k$, for a given point $z \in \mathbb{R}^k$ we may rotate so that

$$z = (z, 0) \in \mathbb{R}^1 \times \mathbb{R}^{k-1} = \mathbb{R}^k.$$

Define $h_t(x, y) = (x + t\sigma(y)\rho(x)z, y)$, where $\sigma : \mathbb{R}^{k-1} \longrightarrow \mathbb{R}$, $\sigma(0) = 1$, $\sigma(y) = 0$ if $|y| \geq \delta$. □

Corollary. *Let Y be a connected manifold of* dim > 1. *If $y_1, \cdots, y_n$, $z_1, \cdots, z_n$ are distinct points in Y, then there exists a diffemorphism $h : Y \longrightarrow Y$, isotopic to the identity, with $h(y_i) = z_i$, $i = 1, 2, \cdots, n$. Moreover, the isotopy may be taken to be compactly supported.*

11.7. Exercises

1. In $\mathbb{R}^2 - \{0\}$, show that

(1) $d\arg(z) = \frac{-ydx}{x^2+y^2} + \frac{xdy}{x^2+y^2}$ where $z = x + iy = re^{i\theta}$ and $\theta = \arg(z)$.

(2) $\int_{S^1} \frac{1}{z} dz = 2\pi i$.

(3) $\int_{S^1} \frac{f(z)}{z} dz = f(0) \cdot 2\pi i$.

(4) if $f : S^1 \longrightarrow S^1$ is defined by $f(z) = z^n$, and ω a 1-form on S^1, then

$$\int_{S^1} f^*\omega = n \int_{S^1} \omega.$$

2. Solve the followings.

(1) Construct a compactly supported n-form on $\mathbb{R}^n$ whose integral is nonzero.

(2) Construct an n-form ω on S^n with nonzero integral.

(3) A closed n-form ω on S^n is exact if and only if $\int_{S^n} \omega = 0$.

(4) The integral $\int_{S^n} : H^n(S^n) \longrightarrow \mathbb{R}$ is an isomorphism.

Chapter 12

Simplicial Spaces

12.1 Simplicial Homology

12.1.1 Simplicial Complex

Definition 12.1.1. (1) A set $\{v_0, \cdots, v_k\} \subset \mathbb{R}^n$ is **convex-independent** if the set $\{v_1 - v_0, \cdots, v_k - v_0\}$ is independent.

(2) $[v_0, v_1, \cdots, v_k] = \left\{ \sum_{i=0}^k a_i v_i \,|\, a_i \geq 0, \ 1 = \sum_{i=0}^k a_i \right\}$ is called a **closed k-simplex**, which is

(i) the smallest convex set containing $\{v_0, \cdots, v_k\}$.

(ii) k is called the **dimension** of the simplex.

(iii) $(a_0, \cdots, a_k)$ is called the **barycentric coordinate** of

$$v = \sum a_i v_i \in [v_0, \cdots, v_k].$$

(iv) $(v_0, \cdots, v_k) = \{v \in [v_0, \cdots, v_k] \,|\, a_i > 0, \ i = 0, \cdots, k\}$ is called an **open simplex**.

(3) Let $[s] := [v_0, \cdots, v_k]$. Then

(i) $v_0, \cdots, v_k$ is called **vertices** of $[s]$.

(ii) the **closed faces** of $[s]$ are closed simplices $[v_{j_0}, \cdots, v_{j_n}]$ where $\{j_0, \cdots, j_n\}$ is a nonempty subset of $\{0, \cdots, k\}$.

(iii) The **open faces** of $[s]$ is the open simplices $(v_{j_0}, \cdots, v_{j_n})$.

Remark 12.1.2. (1) A vertex is a 0-dimensional closed simplex which is a closed and open face.

(2) The open simplex (s) is the interior of the closed simplex $[s]$ which is the closure of (s).

Definition 12.1.3. A **simplicial complex** $\mathcal{K}$ is a finite set of open simplices in some $\mathbb{R}^n$ such that

(1) if $(s) \in \mathcal{K}$, then all open faces of $[s] \in \mathcal{K}$,

(2) (s_1), $(s_2) \in \mathcal{K}$ and $(s_1) \cap (s_2) \neq \emptyset$, then $(s_1) = (s_2)$.

The **dimension** of $\mathcal{K}$ is the maximum dimension of the simplices in $\mathcal{K}$.

Remark 12.1.4. (1) If $\mathcal{K}$ is a simplicial complex, let

$$[\mathcal{K}] = \bigcup_{(s)\in\mathcal{K}} (s) = \bigcup_{(s)\in\mathcal{K}} [s].$$

(2) If $[s]$ is a closed simplex, the collection s of its open faces is a simplicial complex.

(3) The **r-skeleton** $\mathcal{K}^r$ of $\mathcal{K}$ is the collection

$$\mathcal{K}^r = \big\{(s) \in \mathcal{K} \,|\, \dim s \leq r\big\}.$$

(4) if v is a vertex of $\mathcal{K}$, then the star of v is

$$\mathrm{st}(v) = \bigcup_{\substack{v\in[s] \\ (s)\in\mathcal{K}}} (s).$$

Thus $\mathcal{K} = \bigcup_{v\in\mathcal{K}^0} \mathrm{st}(v)$ is an open covering of K.

12.1.2 Simplicial Homology

Definition 12.1.5. Let s be an ℓ-simplex with vertices $v_0, \cdots, v_\ell$. Two ordering $(v_{j_0}, \cdots, v_{j_\ell})$, $(v_{k_0}, \cdots, v_{k_\ell})$ of vertices of s are **equivalent** if $(k_0, \cdots, k_\ell)$ is an even permutation of $(j_0, \cdots, j_\ell)$.
This is an equivalence relation for $\ell \geq 1$, it partitions the orderings of $v_0, \cdots, v_\ell$ into two equivalence classes.

An **oriented simplex** is a simplex s together with a choice of one of these equivalence classes. If $v_0, \cdots, v_\ell$ are the vertices of s, the oriented simplex determined by the ordering $(v_0, \cdots, v_\ell)$ will be denoted by $< v_0, \cdots, v_\ell >$.

Example 12.1.6. (1) 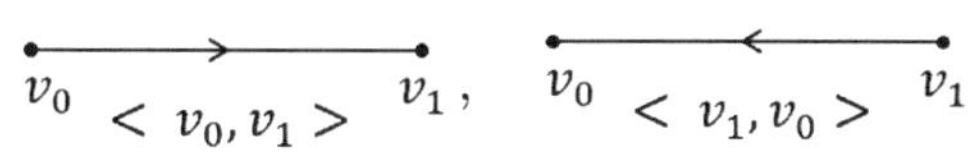

(2) 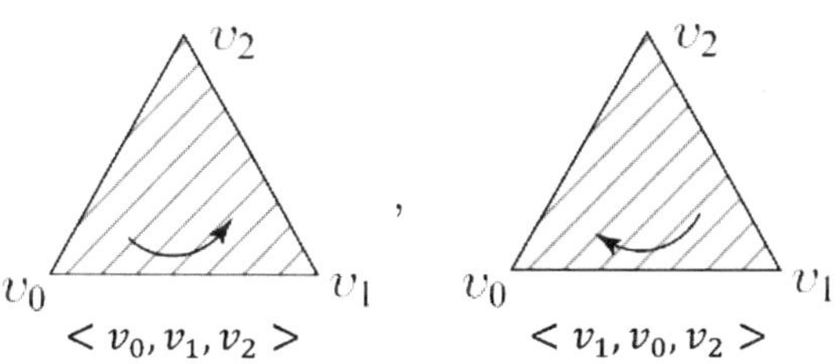

(3) Each ℓ-simplex s lies in an ℓ-dimensional plane in $\mathbb{R}^n$. Orienting s by $< v_0, \cdots, v_\ell >$ is equal to orienting the ℓ-plane containing s by the ordered basis $\{v_1 - v_0, \cdots, v_\ell - v_0\}$.

Definition 12.1.7. Let $\mathcal{K}$ be a simplicial complex. Let $C_\ell(\mathcal{K}, \mathbb{R})$ denote the factor space of the vector space generated by all oriented ℓ-simplices of $\mathcal{K}$, modulo the subspace generated by all elements of the form

$$< v_0, v_1, v_2, \cdots, v_\ell > + < v_1, v_0, v_2, \cdots, v_\ell > .$$

Then $C_\ell(\mathcal{K}, \mathbb{R})$ is called the **space of ℓ-chains** of $\mathcal{K}$ with real coefficients.

A **typical element** of this space is of the form

$$\sum_{s:\text{ an } \ell-\text{simplex in } \mathcal{K}} r_s < s >, \quad r_s \in \mathbb{R}$$

where for each ℓ-simplex s, $< s >$ is s with some fixed orientation of s, and where s with the opposite orientation is identified with $- < s >$.

Definition 12.1.8. Let $< s >=< v_0, \cdots, v_{\ell+1} >$ be an oriented $(\ell+1)$-simplex. Then the **boundary** $\partial < s >$ of $< s >$ is a ℓ-chain defined by

$$\partial < s >= \sum_{i=0}^{\ell+1} (-1)^{\ell} < v_0, \cdots, \hat{v_i}, \cdots, v_{\ell+1} >$$

where ˆ over a symbol means that symbol is deleted.

Example 12.1.9. (1) $\partial < v_0, v_1 >=< v_1 > - < v_0 >$

(2) $$\begin{aligned}\partial < v_0, v_1, v_2 > &=< v_1, v_2 > - < v_0, v_2 > + < v_0, v_1 > \\ &=< v_0, v_1 > + < v_1, v_2 > + < v_2, v_0 >\end{aligned}$$

Definition 12.1.10. Let $\mathcal{K}$ be a simplicial complex. The boundary map

$$\partial : C_{\ell+1}(\mathcal{K}, \mathbb{R}) \longrightarrow C_{\ell}(\mathcal{K}, \mathbb{R})$$

is the **linear map** defined by

$$\partial\big(\sum r_s < s > \big) = \sum r_s \partial < s > .$$

Lemma 12.1.11. *The boundary maps*

$$\cdots \longrightarrow C_{\ell+1}(\mathcal{K}, \mathbb{R}) \xrightarrow{\partial} C_{\ell}(\mathcal{K}, \mathbb{R}) \xrightarrow{\partial} C_{\ell-1}(\mathcal{K}, \mathbb{R}) \xrightarrow{\partial} \cdots$$

satisfies $\partial \circ \partial = 0$.

Proof. $\partial\big(\partial < v_0, \cdots, v_{\ell+1} > \big)$

$$
\begin{aligned}
&= \partial\Big[\sum_{j=0}^{\ell+1}(-1)^j < v_0, \cdots, \widehat{v_j}, \cdots, v_{\ell+1} > \Big] \\
&= \sum_{j=0}^{\ell+1}(-1)^j \partial < v_0, \cdots, \widehat{v_j}, \cdots, v_{\ell+1} >
\end{aligned}
$$

$$
\begin{aligned}
&= \sum_{j=0}^{\ell+1}(-1)^j \Big[\sum_{i=0}^{j-1}(-1)^i < v_0, \cdots, \hat{v}_i, \cdots, \hat{v}_j, \cdots, v_{\ell+1} > \\
&\qquad + \sum_{i=j+1}^{\ell+1} (-1)^{i-1} < v_0, \cdots, \widehat{v_j}, \cdots, \widehat{v_i}, \cdots, v_{\ell+1} > \Big] \\
&= \sum_{i<j}(-1)^{i+j} < v_0, \cdots, \widehat{v_i}, \cdots, \widehat{v_j}, \cdots, v_{\ell+1} > \\
&\qquad + \sum_{i>j}(-1)^{i+j-1} < v_0, \cdots, \widehat{v_j}, \cdots, \widehat{v_i}, \cdots, v_{\ell+1} > \\
&= \sum_{i<j} \big[(-1)^{i+j} + (-1)^{i+j-1}\big] < v_0, \cdots, \widehat{v_i}, \cdots, \widehat{v_j}, \cdots, v_{\ell+1} > \\
&= 0
\end{aligned}
$$

□

Definition 12.1.12. $Z_\ell(\mathcal{K}, \mathbb{R}) = \{c \in C_\ell(\mathcal{K}, \mathbb{R}) \,|\, \partial c = 0\}$ is called the **ℓ-th space of cycles.** $B_\ell(\mathcal{K}, \mathbb{R}) = \{\partial c \,|\, c \in C_{\ell+1}(\mathcal{K}, \mathbb{R})\}$ is called the **ℓ-th space of boundaries.**

$$H_\ell(\mathcal{K}, \mathbb{R}) = Z_\ell / B_\ell(\mathcal{K}, \mathbb{R})$$

is called the **ℓ-th homology space** of $\mathcal{K}$.

Remark 12.1.13. (1) It turns out that the spaces $H_\ell(\mathcal{K}, \mathbb{R})$ depend only on the topology of $[\mathcal{K}]$.

(2) If $f : [\mathcal{K}] \longrightarrow [L]$ is an homeomorphism, then

$$f_* : H_\ell(\mathcal{K}, \mathbb{R}) \longrightarrow H_\ell(L, \mathbb{R}) \text{ is an isomorphism.}$$

Example 12.1.14. (1) Let $\mathcal{K}$ be the 1-skeleton of a 2-simplex

$=$ 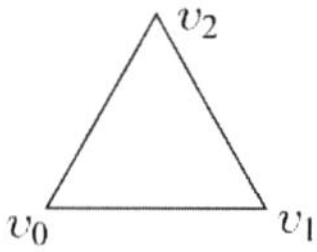

$$= \{v_0, v_1, v_2, (v_0, v_1), (v_1, v_2), (v_2, v_1)\}.$$

Then $C_0(\mathcal{K}, \mathbb{R}) = \mathbb{R} < v_0 > + \mathbb{R} < v_1 > + \mathbb{R} < v_2 >$,
$C_1(\mathcal{K}, \mathbb{R}) = \mathbb{R} < v_0, v_1 > + \mathbb{R} < v_1, v_2 > + \mathbb{R} < v_2, v_0 >$,
$C_\ell(\mathcal{K}, \mathbb{R}) = 0$ if $\ell > 1$.

Let $c_1 = r_1 < v_0, v_1 > + r_2 < v_1, v_2 > + r_3 < v_2, v_0 >$. Then

$$\begin{aligned} \partial c_1 &= r_1\big(< v_1 > - < v_0 > \big) + r_2\big(< v_2 > - < v_1 > \big) \\ &\quad + r_3\big(< v_0 > - < v_2 > \big) \\ &= \big(r_3 - r_1\big) < v_0 > + \big(r_1 - r_2\big) < v_1 > + \big(r_2 - r_3\big) < v_2 > . \end{aligned}$$

Then $c_1 \in Z_1(\mathcal{K}, \mathbb{R}) \iff r_3 - r_1 = 0, \quad r_1 - r_2 = 0, \quad r_2 - r_3 = 0$
$\iff r_1 = r_2 = r_3.$

Thus $Z_1(\mathcal{K}, \mathbb{R}) = \{r(< v_0, v_1 > + < v_1, v_2 > + < v_2, v_0 >) \,|\, r \in \mathbb{R}\}$
$\cong \mathbb{R}.$

Since $C_2(\mathcal{K}, \mathbb{R}) = 0$, $B_1(\mathcal{K}, \mathbb{R}) = 0$, we have

$$H_1(\mathcal{K}, \mathbb{R}) = Z_1(\mathcal{K}, \mathbb{R}) / B_1(\mathcal{K}, \mathbb{R}) \simeq \mathbb{R}.$$

$C_0(\mathcal{K}, \mathbb{R}) = Z_0(\mathcal{K}, \mathbb{R})$.

If $c_0 = \lambda_1(v_0) + \lambda_2(v_1) + \lambda_3(v_2)$, then

$$\begin{aligned} \partial c_1 &= \partial\big(r_1 < v_0, v_1 > + r_2 < v_1, v_2 > + r_3 < v_2, v_0 > \big) \\ &= (r_3 - r_1) < v_0 > + (r_1 - r_2) < v_1 > + (r_2 - r_3) < v_2 > \end{aligned}$$

$$\iff \begin{pmatrix} r_3 - r_1 = \lambda_1 \\ r_3 - r_1 = \lambda_1 \\ r_3 - r_1 = \lambda_1 \end{pmatrix} \iff \lambda_1 + \lambda_2 + \lambda_3 = 0.$$

Thus $B_0(\mathcal{K}, \mathbb{R}) = \{\lambda_1 < v_0 > + \lambda_2 < v_1 > + \lambda_3 < v_2 > \,|\, \lambda_1 + \lambda_2 + \lambda_3 = 0\}$.

Let $\varphi : Z_0(\mathcal{K}, \mathbb{R}) \longrightarrow \mathbb{R}$,

$$\varphi\big(\lambda_1 < v_0 > + \lambda_2 < v_1 > + \lambda_3 < v_3 > \big) = \lambda_1 + \lambda_2 + \lambda_3.$$

Then φ is an epimorphism and $\varphi^{-1}(0) = B_0(\mathcal{K}, \mathbb{R})$. Thus

$$H_0(\mathcal{K}, \mathcal{T}) = Z_0(\mathcal{K}, \mathbb{R})/B_0(\mathcal{K}, \mathbb{R}) \simeq \mathbb{R}.$$

(2) Let $\mathcal{K}$ be the simpicial complex consisting of all the faces of a 2-simplex (v_0, v_1, v_2)

$$= \{(v_0), (v_1), (v_2), (v_0, v_1), (v_1, v_2), (v_2, v_0), (v_0, v_1, v_2)\}.$$

Then $C_0(\mathcal{K}, \mathbb{R}) = \{\lambda_1 < v_0 > +\lambda_2 < v_1 > +\lambda_3 < v_2 > \mid \lambda_1, \lambda_2, \lambda_3 \in \mathbb{R}\}$,
$C_1(\mathcal{K}, \mathbb{R}) = \{r_1 < v_0, v_1 > +r_2 < v_1, v_2 > +r_3 < v_2, v_0 > \mid r_1, r_2, r_3 \in \mathbb{R}\}$,
$C_2(\mathcal{K}, \mathbb{R}) = \{r < v_0, v_1, v_2 > \mid r \in \mathbb{R}\}$.

By (1), we have $H_0(\mathcal{K}, \mathbb{R}) \simeq \mathbb{R}$.

$$Z_1(\mathcal{K}, \mathbb{R}) = \{r(< v_0, v_1 > + < v_1, v_2 > + < v_2, v_0 >) \,|\, r \in \mathbb{R}\}.$$

$$\begin{aligned} B_1(\mathcal{K}, \mathbb{R}) &= \{\partial(r < r_0, r_1, r_2 >) \,|\, r \in \mathbb{R}\} \\ &= \{r\partial < v_0, v_1, v_2 \,|\, r \in \mathbb{R}\} = Z_1(\mathcal{K}, \mathbb{R}). \end{aligned}$$

Thus $H_1(\mathcal{K}, \mathbb{R}) = Z_1(\mathcal{K}, \mathbb{R})/B_1(\mathcal{K}, \mathbb{R}) = \{0\}$.

$$\partial(r < r_0, r_1, r_2 >) = 0 \quad \Longleftrightarrow \quad r = 0,$$

thus $Z_2(\mathcal{K}, \mathbb{R}) = 0$, and hence $H_2(\mathcal{K}, \mathbb{R}) = 0$.

Definition 12.1.15. Let $\mathcal{K}$ be a simplicial complex. Then

(1) the ℓ-th **Betti number** b_ℓ of $\mathcal{K}$ is the integer

$$b_\ell := \dim H_\ell(\mathcal{K}, \mathbb{R}).$$

(2) the **Euler characteristic** $\chi(\mathcal{K})$ of $\mathcal{K}$ is the integer

$$\chi(\mathcal{K}) := \sum_{\ell=0} (-1)^\ell b_\ell.$$

Theorem 12.1.16. *Let $\mathcal{K}$ be a simplical complex and α_ℓ the number of ℓ-simplices in $\mathcal{K}$. Then*

$$\chi(\mathcal{K}) = \sum_{\ell=0} (-1)^\ell \alpha_\ell.$$

Proof. For each ℓ, $0 \leq \ell \leq \dim \mathcal{K}$, consider the linear map

$$\partial_\ell : C_\ell(\mathcal{K}, \mathbb{R}) \longrightarrow C_{\ell-1}(\mathcal{K}, \mathbb{R}).$$

Then $\alpha_\ell = \dim C_\ell(\mathcal{K}, \mathbb{R})$

$= \dim \mathrm{Ker}\partial_\ell + \dim \mathrm{Im}\partial_\ell = \dim Z_\ell(\mathcal{K}, \mathbb{R}) + \dim B_{\ell-1}(\mathcal{K}, \mathbb{R})$.

Moreover, $b_\ell = \dim H_\ell(\mathcal{K}, \mathbb{R}) = \dim(Z_\ell(\mathcal{K}, \mathbb{R})/B_\ell(\mathcal{K}, \mathbb{R}))$

$= \dim Z_\ell(\mathcal{K}, \mathbb{R}) - \dim B_\ell(\mathcal{K}, \mathbb{R})$.

Thus

$$\begin{aligned}
\chi(\mathcal{K}) &= \sum_{\ell=0}^{\dim \mathcal{K}} (-1)^\ell b_\ell \\
&= \sum_{\ell=0}^{\dim \mathcal{K}} (-1)^\ell \big[\dim Z_\ell(\mathcal{K}, \mathbb{R}) - \dim \mathrm{B}_\ell(\mathcal{K}, \mathbb{R})\big] \\
&= \sum_{\ell=0}^{\dim \mathcal{K}} (-1)^\ell \dim Z_\ell(\mathcal{K}, \mathbb{R}) + \sum_{\ell=0}^{\dim \mathcal{K}} (-1)^{\ell+1} \dim B_\ell(\mathcal{K}, \mathbb{R}) \\
&= \sum_{\ell=0}^{\dim \mathcal{K}} (-1)^\ell \dim Z_\ell(\mathcal{K}, \mathbb{R}) + \sum_{\ell=1}^{\dim \mathcal{K}} (-1)^\ell \dim B_{\ell-1}(\mathcal{K}, \mathbb{R}) \\
&= \sum_{\ell=0}^{\dim \mathcal{K}} (-1)^\ell \big[\dim Z_\ell(\mathcal{K}, \mathbb{R}) + \dim B_{\ell-1}(\mathcal{K}, \mathbb{R})\big] \\
&= \sum_{\ell=0}^{\dim \mathcal{K}} (-1)^\ell \alpha_\ell, \quad \text{since } \dim B_0 = \dim B_{\dim \mathcal{K}}.
\end{aligned}$$

□

Example 12.1.17. If $[\mathcal{K}]$ is a connected compact orientable 2-dimensional manifold, then

$$\chi(\mathcal{K}) = b_0 - b_1 + b_2 = 2 - b_1, \quad \text{since } b_0 = b_2 = 1.$$

So $b_1 = 2\chi(\mathcal{K})$, b_1 is always even, say $b_1 = 2g$.

Thus $\chi(\mathcal{K}) = 2 - 2g$, g is called the **genus** of $\mathcal{K}$.

12.1. Exercises

1. Let K be the 2-skeleton of a 3-simplex $[v_0, v_1, v_2, v_3]$.

(1) Find the space $C_\ell(K, \mathbb{R})$ of ℓ-chains of K with real coefficients.

(2) Find the boundary map $\partial : C_{\ell+1}(K, \mathbb{R}) \longrightarrow C_\ell(K, \mathbb{R})$.

(3) Find the space $Z_\ell(K, \mathbb{R})$ of ℓ-cycles of K.

(4) Find the space $B_\ell(K, \mathbb{R})$ of ℓ-boundaries of K.

(5) Find the simplicial homology group of K, $H_\ell(K, \mathbb{R})$.

(6) Find the Euler characteristic $\chi(K)$ of K.

12.2 Simplicial Cohomology

To compare the homology of simplicial complexes with de Rham cohomology theory, it is convenient to define a simplicial cohomology theory.

Definition 12.2.1. Let $\mathcal{K}$ be a simplicial complex.
For $0 \leq \ell \leq \dim \mathcal{K}$, let the ℓ-th space of cochains in $\mathcal{K}$

$$C^\ell(\mathcal{K}) = C_\ell(\mathcal{K}, \mathbb{R})^*.$$

Define a **coboundary operator** $\partial^* : C^\ell(\mathcal{K}) \longrightarrow C^{\ell+1}(\mathcal{K})$ by the adjoint of the boundary operator

$$\partial : C_{\ell+1}(\mathcal{K}, \mathbb{R}) \longrightarrow C_\ell(\mathcal{K}, \mathbb{R}), \quad [\partial^*(\varphi)](c) = \varphi(\partial c),$$

for $\varphi \in C^\ell(\mathcal{K})$ and $c \in C_{\ell+1}(\mathcal{K}, \mathbb{R})$. Then we have a sequence

$$\cdots \longrightarrow C^{\ell-1}(\mathcal{K}) \xrightarrow{\partial^*} C^\ell(\mathcal{K}) \xrightarrow{\partial^*} C^{\ell+1}(\mathcal{K}) \longrightarrow \cdots.$$

Since $\partial \circ \partial = 0$, $\partial^* \circ \partial^* = 0$.

Definition 12.2.2. (1) $Z^\ell(\mathcal{K}) := \{\varphi \in C^\ell(\mathcal{K}) \,|\, \partial^*(\varphi) = 0\}$ is called the **group of cocycles**.

(2) $B^\ell(\mathcal{K}) := \{\partial^* \varphi \,|\, \varphi \in C^{\ell-1}(\mathcal{K})\}$ is called the **group of coboundaries**.

(3) $H^\ell(\mathcal{K}) := \frac{Z^\ell(\mathcal{K})}{B^\ell(\mathcal{K})}$ is called the **ℓ-th cohomology group** of $\mathcal{K}$.

Exhibit an explicit formula of the coboundary operator ∂^*. For each oriented ℓ-simplex $< s >$ of $\mathcal{K}$. Define $\varphi_{<s>} \in C^\ell(\mathcal{K})$ by

$$\varphi_{<s>} < t >= \begin{cases} 1 & \text{if } < s >=< t > \\ -1 & \text{if } < s >= - < t > \\ 0 & \text{if } s \neq t. \end{cases}$$

If $\{< s_1 >, \cdots, < s_m >\}$ is a basis for $C_\ell(\mathcal{K}, \mathbb{R})$, then $\{\varphi_{<s_1>}, \cdots, \varphi_{<s_m>}\}$ is the dual basis for $C^\ell(\mathcal{K})$.

Theorem 12.2.3. (1) *$H^\ell(\mathcal{K})$ is isomorphic to $H_\ell(\mathcal{K},\mathbb{R})^*$.*

(2)

$$\partial^*\varphi_{\langle v_0,\cdots,v_\ell\rangle} = \sum_v{}' \varphi_{\langle v,v_0,\cdots,v_\ell\rangle}$$

where $\sum_v{}'$ denotes the sum over all vectors $v \in \mathcal{K}$ such that $(v, v_0, \cdots, v_\ell)$ is an $(\ell+1)$-simplex in $\mathcal{K}$.

Proof. (1) It is clear by definition.

(2) For an oriented $(\ell+1)$-simplex $\langle t\rangle = \langle w_0, w_1, \cdots, w_{\ell+1}\rangle$ of $\mathcal{K}$, and by setting $\langle s\rangle = \langle v_0, v_1, \cdots, v_\ell\rangle$,

$$\begin{aligned}(\partial^*\varphi_{\langle x\rangle})(\langle t\rangle) &= \varphi_{\langle s\rangle}(\partial \langle t\rangle)\\ &= \varphi_{\langle s\rangle}\Big(\sum_{i=0}^{\ell+1}(-1)^i \langle w_0,\cdots,\widehat{w_i},\cdots,w_{\ell+1}\rangle\Big)\\ &= \sum_{i=0}^{\ell+1}(-1)^i\varphi_{\langle s\rangle}(\langle w_0,\cdots,\widehat{w_i},\cdots,w_{\ell+1}\rangle)\end{aligned}$$

Each term of this sum is zero unless for some j,

$$(w_0, \cdots, \widehat{w_j}, \cdots, w_{\ell+1}) = (s),$$

that is, unless (s) is a face of (t).

If (s) is a face of (t), then $(t) = (v, v_0, \cdots, v_\ell)$ for some vertex $v \in \mathcal{K}$, in which case either

(1) $\langle t\rangle = \langle v, v_0, \cdots, v_\ell\rangle$ and $\partial^*\varphi_{\langle s\rangle}(\langle t\rangle) = 1$, or

(2) $\langle t\rangle = -\langle v, v_0, \cdots, v_\ell\rangle$ and $\partial^*\varphi_{\langle s\rangle}(\langle t\rangle) = -1$.

Thus we have

$$\begin{aligned}(\partial^*\varphi_{\langle s\rangle})(\langle t\rangle) &= \begin{cases} 1 & \text{if } \langle t\rangle = \langle v, v_0, \cdots, v_\ell\rangle \\ -1 & \text{if } \langle t\rangle = -\langle v, v_0, \cdots, v_\ell\rangle \\ 0 & \text{in all other cases}\end{cases}\\ &= \Big(\sum_v{}' \varphi_{\langle v,v_0,\cdots,v_\ell\rangle}\Big)(\langle t\rangle).\end{aligned}$$

Since this holds for arbitrary $< t >$, the formula is established.

□

12.2. Exercises

1. Let $\mathcal{K}$ be the 2-skeleton of a 3-simple $[v_0, v_1, v_2, v_3]$.

(1) Find the space $C^\ell(\mathcal{K}, \mathbb{R})$ of ℓ-cochains of $\mathcal{K}$ with real coefficients.

(2) Find the coboundary map $\partial^* : C^\ell(\mathcal{K}, \mathbb{R}) \longrightarrow C^{\ell+1}(\mathcal{K}, \mathbb{R})$.

(3) Find the space $Z^\ell(\mathcal{K}, \mathbb{R})$ of ℓ-cocycles of $\mathcal{K}$.

(4) Find the space $B^\ell(\mathcal{K}, \mathbb{R})$ of ℓ-coboundaries of $\mathcal{K}$.

(5) Find the simpicial cohomology group $H^\ell(\mathcal{K}, \mathbb{R})$ of $\mathcal{K}$ with real coefficients.

12.3 De Rham's Theorem

12.3.1 Smooth Triangulation

Definition 12.3.1. A **smoothly triangulated manifold** is a triple $(X, \mathcal{K}, h)$, where X is a smooth manifold, $\mathcal{K}$ is a simplicial complex, and $h : [k] \longrightarrow X$ is a homeomorphism such that for each simplex s of $\mathcal{K}$, the map $h|_{[s]} : [s] \longrightarrow X$ has an extension h_s to a nbd U of $[s]$ in the plane of $[s]$ such that $h_s : U \longrightarrow X$ is a smooth submanifold.

Remark 12.3.2. If $\dim X = n$, we need only require that this last condition be satisfied for each n-simplex of $\mathcal{K}$, since every simplex of $\mathcal{K}$ is a face of an n-simplex and restrictions of smooth maps to submanifolds are smooth.

Example 12.3.3. Let $X = S^n$. Let $\mathcal{K}$ be the n-skeleton of an $(n+1)$-simplex circumscribed about S^n.

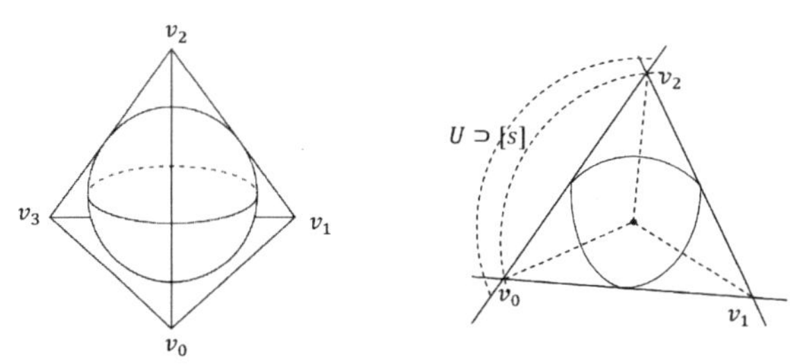

$$\begin{array}{ccc} [s] & \xrightarrow{h|_{[s]}} & S^n \\ \big\downarrow & \nearrow_{h_s} & \\ U & & \end{array}$$

where h_s is a smooth extension of $h|_{[s]}$. $h : [k] \longrightarrow S^n$ is a homeomorphism.

Theorem 12.3.4. (1) *Every compact smooth manifold can be smoothly triangulated.*

(2) *For smoothly triangulated manifolds* $(X, \mathcal{K}, h)$, *the de Rham cohomology of* X *is isomorphic to the simplicial cohomology of* K.

Definition 12.3.5. Let $\mathcal{K}$ be a simplicial complex and let $v_1, \cdots, v_m$ denote the vertices of $\mathcal{K}$, and $x \in [k]$. For $j \in \{1, 2, \cdots, m\}$, the **j-th barycentric coordinate** $b_j(x)$ of x is defined as follows.

If $x \notin \mathrm{st}(v_j)$, then $b_j(x) = 0$.

If $x \in \mathrm{st}(v_j)$, then $x \in (s)$ for some simplex s having v_j as a vertex, and $b_j(x)$ is the barycentric coordinate of x in s relative to the vertex v_j.

Lemma 12.3.6. (1) $b_j : [k] \longrightarrow \mathbb{R}$ is a continuous map.

(2) $b_j(x) \geq 0$ and $\sum_{i=1}^{m} b_j(x) = 1$ for each $x \in [k]$.

(3) $x = \sum_{j=1}^{m} b_j(x) v_j$.

(4) $b_{j_0}(x) \neq 0$, $b_{j_1}(x) \neq 0$, $\cdots$, $b_{j_\ell}(x) \neq 0$ for some $x \in [k]$ iff $v_{j_0}, \cdots, v_{j_\ell}$ are the vertices of an ℓ-simplex of $\mathcal{K}$ containing x.

Definition 12.3.7. Let $\mathcal{K}$ be a simplicial complex, $s \in \mathcal{K}$. The **star** of s is given by

$$\mathrm{st}(s) = \bigcup_{(s):\ \text{a face of } [t]} (t).$$

Lemma 12.3.8. (1) *If* $s = v$ *is a* 0*-simplex,* $\mathrm{st}(s) = \mathrm{st}(v)$.

(2) $\mathrm{st}(s)$ *is an open set in* $[k]$.

(3) *If* $(s) = (v_{j_0}, \cdots, v_{j_\ell})$ *and* $x \in [k]$, *then* $x \in \mathrm{st}(s)$ *iff* $b_{ji}(x) \neq 0$ *for all* $i \in \{0, \cdots, \ell\}$.

(4) *If* $(s) = (v_{j_0}, \cdots, v_{j_\ell})$, *then*

$$[k] - \mathrm{st}(s) = \Big[x \in [k] \,|\, b_{ji}(x) = 0 \quad \textit{for some } i \in \{1, \cdots, \ell\}\Big].$$

(5) *If* s_1 *and* s *are* ℓ*-simplices in* $\mathcal{K}$ *with* $s_1 \neq s$, *then* $[s_1] \subset [k] - \mathrm{st}(s)$.

12.3.2 De Rham Isomorphism

Let $(X, \mathcal{K}, h)$ be a smoothly triangulated manifold. For each ℓ, we define a sequence of linear maps

$$\int_\ell : C^\infty(X, \bigwedge{}^\ell(X)) \longrightarrow C^\ell(\mathcal{K})$$

for each $\omega \in C^\infty(X, \wedge^\ell(X))$, an oriented ℓ-simplex $< s > \in C_\ell(\mathcal{K})$, consider a smooth function $h_s : U \longrightarrow X$, where U is an open set in the plane of $[s]$, i.e., in an ℓ-dimensional Euclidean space. We define

$$\int_\ell (\omega)(< s >) := \int_{<s>} h_s^*(\omega).$$

In other word, let $(r_1, \cdots, r_\ell)$ denote the coordinates in the plane of $[s]$ consistent with the orientation of $< s >$.

Let $< s > = < v_0, \cdots, v_\ell >$ and let $(r_1, \cdots, r_\ell)$ be coordinates relative to the ordered basis $\{v_1 - v_0, \cdots, v_\ell - v_0\}$. Then

$$h_s^*(\omega) = g\, dr_1 \wedge \cdots \wedge dr_\ell,$$

for some continuous map g on U, and

$$\int_\ell (\omega)(< s >) := \int_{[s]} g\, dr_1 \wedge \cdots \wedge dr_\ell.$$

Remark 12.3.9. This integral is independent of the h, i.e., it depends only on the set $h([s])$ and its orientation.

Theorem 12.3.10 (Stokes' Theorem).

$$\partial^* \circ \int_\ell = \int_{\ell+1} \circ\, d,$$

$$\begin{array}{ccc} C^\infty(X, \wedge^\ell X) & \xrightarrow{d} & C^\infty(X, \wedge^{\ell+1} X) \\ \downarrow \int_\ell & & \downarrow \int_{\ell+1} \\ C^\ell(\mathcal{K}) & \xrightarrow{\partial^*} & C^{\ell+1}(\mathcal{K}) \end{array}$$

Proof. For an ℓ-form ω and $(\ell+1)$-simplex $< s >$,

$$\begin{aligned} \Big[\big(\int_{\ell+1} \circ\, d\big)(w)\Big](< s >) &= \int_{<s>} h_s^*(d\omega) \\ &= \int_{<s>} d(h_s^*\omega) = \int_{\partial<s>} h_s^*\omega \\ &= \int_\ell (\omega)(\partial < s >) = \Big[\big(\partial^* \circ \int_\ell\big)(\omega)\Big](< s >). \end{aligned}$$

□

Thus $\int_\ell(Z^\ell(X,d)) \subset Z^\ell(\mathcal{K})$, since $d\omega = 0$, $\partial^*(\int_\ell \omega) = 0$. And $\int_\ell(B^\ell(X,d) \subset B^\ell(\mathcal{K})$, since $\int_\ell \omega = \int_\ell dc = \partial^* \int_{\ell+1} c \in \operatorname{Im}\partial^*$. Thus $\int_\ell$ induces a homomorphism

$$\widetilde{\int_\ell} : H^\ell(X,d) \longrightarrow H^\ell(\mathcal{K}).$$

Theorem 12.3.11 (de Rham's Theorem). *Let $(X, \mathcal{K}, h)$ be a smoothly triangulated manifold. Then*

$$\widetilde{\int_\ell} : H^\ell(X,d) \longrightarrow H^\ell(\mathcal{K})$$

is an isomorphism for each ℓ, $0 \le \ell \le \dim X$.

Lemma 12.3.12. *There exists a sequence of linear maps*

$$\alpha_\ell : C^\ell(\mathcal{K}) \longrightarrow C^\infty(X, \wedge^\ell(X)), \quad 0 \le \ell \le \dim X$$

such that

(1) $d \circ \alpha_\ell = \alpha_{\ell+1} \circ \partial^*$,

(2) $\int_\ell \circ \alpha_\ell =$ identity,

(3) *If* c^0 *denotes the* 0*-cochain such that* $c^0(v) = 1$ *for each vertex* v *in* $\mathcal{K}$*, then*

$$\alpha_0(c^0) = 1,$$

i.e., $\alpha_0(c^0) : X \longrightarrow \mathbb{R}$ *is the constant map* $\alpha_0(c^0) = 1$.

(4) *If* $< s >$ *is an oriented* ℓ*-simplex of* $\mathcal{K}$*, then the* ℓ*-form* $\alpha_\ell(\varphi_{<s>})$ *is identically zero in a nbd of* $X - \mathrm{st}(s)$.

Lemma 12.3.13. *Let* ω *be a closed* ℓ*-form on* X.
Suppose that $\int_\ell(\omega) = \partial^* c$ *for some* $c \in C^{\ell-1}(\mathcal{K})$.
Then there exists an $(\ell - 1)$*-form* τ *on* X *such that*

$$\int_{\ell-1} \tau = c \quad \text{and} \quad d\tau = \omega.$$

Proof of Theorem 12.3.11. Under assuming Lemmas 12.3.12 and 12.3.13.
For given $z \in Z^\ell(\mathcal{K})$, let $\omega = \alpha_\ell(z)$. Then

$$d\omega = d \circ \alpha_\ell(z) = \alpha_{\ell+1}\partial^*(z) = \alpha_{\ell+1}(0) = 0.$$

Furthermore, $\int_\ell \omega = \int_\ell \circ \alpha_\ell(z) = z$, by Lemma 12.3.12. Thus

$$\int_\ell : Z^\ell(X, d) \longrightarrow Z^\ell(\mathcal{K})$$

is surjective, hence so is $\widetilde{\int_\ell}$. By Lemma 12.3.13, $\widetilde{\int_\ell}$ is injective.

For if $\omega \in Z^\ell(X, d)$ and $\int_\ell(\omega) \in B^\ell(\mathcal{K})$, then there is an $(\ell-1)$-form τ on X such that $d\tau = \omega$, by Lemma 12.3.13. Thus $\mathrm{Ker}\widetilde{\int_\ell} = B^\ell(X, d)$. Therefore $\widetilde{\int_\ell} : H^\ell(X, d) \longrightarrow H^\ell(\mathcal{K})$ is an isomorphism. □

Let $(X, \mathcal{K}, h)$ be a smoothly triangulated manifold. Then we have the following isomorphisms :

$$
\begin{array}{ccccccc}
H^\ell(X,d) & \xrightarrow[\text{Thm 12.3.11}]{\widetilde{\int_\ell}} & H^\ell(\mathcal{K},\mathbb{R}) & \xrightarrow[\text{Thm 12.2.3}]{\sim} & H_\ell(\mathcal{K},\mathbb{R})^* & \xrightarrow{\text{dual}} & H_\ell(\mathcal{K},\mathbb{R}) \\
 & & \| & & \| & & \| \\
 & & H^\ell(X,\mathbb{R}) & & H_\ell(X,\mathbb{R})^* & & H^\ell(X,\mathbb{R})
\end{array}
$$

and

$$\Big(\widetilde{\int_\ell}(\omega)\Big)\sigma = \int_\sigma \omega =< \omega, \sigma > \in \mathbb{R}.$$

Thus the pairing

$$<\,,\,>:\ H^\ell(X,d) \times H_\ell(X,\mathbb{R}) \longrightarrow \mathbb{R}$$

is given by

$$(\omega,\sigma) \mapsto < \omega, \sigma >= \int_\sigma \omega$$

is a non-degenerate bilinear pairing.

Remark 12.3.14. (1) De Rham's theorem shows that the simplicial cohomology groups with coefficients in $\mathbb{R}$ of a smoothly triangulated manifold $(X,\mathcal{K},h)$ are isomorphic to the de Rham cohomology group of X. In particular, these groups are independent of the triangulation $(\mathcal{K},h)$ of X.

(2) The direct sum $\sum_{\ell=0}^{\dim X} H^\ell(X,d)$ has a structure of an associative algebra :

$$\sum_{\ell=0}^{\dim X} C^\infty(X,\wedge^\ell(X)) \text{ is an associative algebra under } \wedge.$$

$\sum_{\ell=0}^{\dim X} Z^\ell(X,d)$ is a subalgebra, since if $d\omega = 0 = d\tau$, then $d(\omega\wedge\tau) = d\omega\wedge\tau \pm \omega\wedge d\tau = 0$.

$\sum_{\ell=0}^{\dim X} B^\ell(X,d)$ is an ideal in $\sum Z^\ell(X,d)$, since if $\omega = d\mu$ and $d\tau = 0$, then $\omega\wedge\tau = d(\mu\wedge\tau)$. Hence

$$
\begin{aligned}
\sum_\ell H^\ell(X,d) &= \sum_\ell (Z^\ell(X,d)/B^\ell(X,d)) \\
&\cong (\sum_\ell Z^\ell(X,d))/(\sum_\ell B^\ell(X,d))
\end{aligned}
$$

is also an associative algebra, called the **de Rham cohomology ring** of X. Thus we have

$$\int : \Big(\sum_\ell H^\ell(X,d), +, \wedge\Big) \longrightarrow (\sum_\ell H^\ell(\mathcal{K},\mathbb{R}), +, \cup)$$

is an algebra isomorphism.

(3) $H^\ell(S^n, d) = \begin{cases} 0, & 0 < \ell < n, \\ \mathbb{R}, & \ell = 0,\ n. \end{cases}$

Hint: $[S^n]$ is the n-skeleton of an $(n+1)$-simplex s.

12.3. Exercises

1. Construct the n-skeleton of an $(n+1)$-simplex circumscribed about S^n.

2. Prove Lemma 12.3.6.

3. Prove Lemma 12.3.8.

4. Prove Lemma 12.3.12.

5. Prove Lemma 12.3.13.

6. Prove that the pairing for a smooth compact manifold X

$$< , >: H^\ell_{\text{de Rham}}(X,\mathbb{R}) \times H_\ell(X,\mathbb{R}) \longrightarrow \mathbb{R}$$

$$<\omega, \sigma> = \int_\sigma \omega$$

is a nondegenerate bilinear pairing.

7. Let X be a compact smooth manifold of dimension n with a smooth triangulation $\mathcal{K}$. Then show the map

$$\int : \Big(\bigoplus_{i=0}^n H^i_{\text{de Rham}}(X,\mathbb{R}), +, \wedge\Big) \longrightarrow \Big(\bigoplus_{i=0}^n H^i(\mathcal{K},\mathbb{R}), +, \cup\Big)$$

is an algebra isomorphism.

Bibliography

[1] Y. S. Cho, *Algebraic Topology.* Kyung Moon Press, 2010.

[2] Y. S. Cho, *Differential Topology on Manifolds.* Arche, 2002.

[3] Y. S. Cho, *Differentiable Manifolds.* Kyowoo Press, 2019.

[4] Y. S. Cho, *Index Theory.* Kyung Moon Press, 2012.

[5] Y. S. Cho, *Sheaf Theory.* Kyung Moon Press, 2011.

[6] J. L. Kelley, *General Topology.* Toronto-New York-London: D. Van Nostrand, 1955.

[7] V. Guillemin and A. Pollack, *Differential Topology.* Englewood Cliffs, N.J.: Prentice-Hall, 1974.

[8] J. Milnor, *Topology from the Differential Viewpoint.* Charlottesville, Va.: The University Press of Virginia, 1965. Based on notes by David W. Weaver.

[9] J. R. Munkres, *Topology.* Inc., Upper Saddle River, NJ: Prentice Hall, 2nd ed., 2000.

[10] I. M. Singer and J. A. Thorpe, *Lecture Notes on Elementary Topology and Geometry.* Glenview, Ill.: Scott, Foresman and Co., 1967.

[11] F. W. Warner, *Foundations of Differentiable Manifolds and Lie Groups.* Glenview, Ill.-London: Scott, Foresman and Co., 1971.

[12] R. O. Wells, *Differential Analysis on Complex Manifolds.* New York-Berlin: Springer-Verlag, 2nd ed., 1980. Graduate Texts in Mathematics, 65.

Index